DIRTY CHICK

DIRTY CHICK

ADVENTURES OF AN UNLIKELY FARMER

ANTONIA MURPHY

GOTHAM BOOKS

GOTHAM BOOKS
Published by the Penguin Group
A division of Penguin Random House LLC
1745 Broadway, New York, NY 10019
penguinrandomhouse.com

USA | Canada | UK | Ireland | Australia | New Zealand | India | South Africa | China
penguin.com
A Penguin Random House Company

Copyright © 2015 by Antonia Murphy
Penguin Random House values and supports copyright. Copyright fuels creativity, encourages diverse voices, promotes free speech and creates a vibrant culture. Thank you for buying an authorized edition of this book and for complying with copyright laws by not reproducing, scanning, or distributing any part of it in any form without permission. You are supporting writers and allowing Penguin Random House to continue to publish books for every reader. Please note that no part of this book may be used or reproduced in any manner for the purpose of training artificial intelligence technologies or systems.

Gotham Books and the skyscraper logo are trademarks of Penguin Group (USA) LLC.

LIBRARY OF CONGRESS CATALOGING-IN-PUBLICATION DATA

Murphy, Antonia, 1975–
   Dirty chick : adventures of an unlikely farmer / Antonia Murphy.
     pages cm
   ISBN 978-1-59240-905-1 (hc)
   1. Farm life—California—Anecdotes. 2. Women farmers—California—Anecdotes.
3. Murphy, Antonia, 1975– I. Title.
   S521.5.C2M87 2015
   636.09794—dc23   2014025087

Trade Paperback ISBN: 9781592409549

Set in Adobe Caslon Pro Regular
Designed by Elke Sigal

While the author has made every effort to provide accurate telephone numbers, Internet addresses, and other contact information at the time of publication, neither the publisher nor the author assumes any responsibility for errors or for changes that occur after publication. Further, the publisher does not have any control over and does not assume any responsibility for author or third-party websites or their content.

The authorized representative in the EU for product safety and compliance is Penguin Random House Ireland, Morrison Chambers, 32 Nassau Street, Dublin D02 YH68, Ireland, https://eu-contact.penguin.ie.

For Peter,
my love.

CONTENTS

Acknowledgments · IX

Author's Note · XI

Prologue · XIII

Chapter One   It Started with a Chicken · 1

Chapter Two   Freaky Eggs · 14

Chapter Three   Party Time · 25

Chapter Four   Teddy Bear Camels · 34

Chapter Five   Sheep on a Spit · 53

Chapter Six   Stripper Calves with Satan Tongues · 66

Chapter Seven   Back from Love Mountain · 88

Chapter Eight   Turkey Time · 104

Chapter Nine   The First One Is Free · 115

Chapter Ten	Heavy Breathing ·	133
Chapter Eleven	Spinning and Spitting ·	150
Chapter Twelve	Frog in Mouth ·	164
Chapter Thirteen	The Grab-and-Yank ·	179
Chapter Fourteen	Outnumbered ·	198
Chapter Fifteen	Catch the Big One ·	209
Chapter Sixteen	The Binglee-Doo ·	225
Chapter Seventeen	No Crocodiles Here ·	236
Epilogue · 253		

ACKNOWLEDGMENTS

So many people helped make, shape, and publish *Dirty Chick*. Great big thanks and fierce hugs go out to:

My beautiful blonde shark agent, Elizabeth Evans, and the team at the Jean V. Naggar Literary Agency,

My exacting editors in New York and Melbourne, Brooke Carey and Alaina Gougoulis,

Michael Heyward at Text, who "fell about laughing" at my revolting lamb jokes,

Torre DeRoche and Jillian Lauren, fellow writers who helped me find an agent in the first place,

My mother and father, Anne Stein and Alexis Tellis, who read to me always,

Peter, "the perfect man for me," who laughed at early drafts and watched the kids so I could write,

Rebecca, who helped so much with our family and other animals, and

The people of Purua who have made us feel so welcome in their community.

AUTHOR'S NOTE

This is a work of nonfiction. Some names, nationalities, and medical details have been changed to protect people's privacy, the timeline has been compressed to make for a snappier read, and conversations are paraphrased from memory. As I fleshed out the character of Skin, I blended him with memories of two other men: his close friend and his neighbor.

And I'm sorry to say, all the disgusting animal stories actually happened.

PROLOGUE

As I watched my goat eat her placenta, I was mostly impressed. I did experience some other feelings, such as horror and revulsion, and also a hint of nausea. But Pearl had always been a strict vegan, so her sudden craving for raw meat showed a real taste for adventure. Some vegetarians who broaden their menu choices might spring for an egg or a soup made with chicken stock. But my goat, Pearl, went straight for the autocannibalism. Which was disgusting. And also worth some respect.

The placenta draped gracefully from her hindquarters, a translucent pink train enclosing a network of blue veins. There was a dark red, ropy thing inside, heavy with blood and the color of liver. It was this that Pearl tucked into first, craning her neck to nibble and swallow, a bloody moustache staining the white fur around her mouth. Then one of her new babies bleated, and she whipped her head around in midmunch, flinging the placenta across the grass like a parade banner. She licked the amber membrane from her newborn baby's fleece, and when a trickle of meconium appeared between its legs, she licked that up, too.

I regarded my goat, whose face was smeared with a war paint of baby butt tar and blood. Round the back end, things were worse: it looked like an abattoir was coming out of her ass. *Why am I here?* I found myself wondering. *How did my life come to this?*

The answer was clear. It had started with a chicken.

DIRTY CHICK

DIRTY CHICK

CHAPTER ONE

IT STARTED WITH A CHICKEN

It wasn't even my chicken, actually, but my father's, which I was charged with caring for during the three weeks he went on holiday in France with his wife, Gail. This particular chore came with free use of his Spanish-style hacienda with a swimming pool just north of San Francisco, so I was more than happy to oblige. Besides the handful of chickens, there was an elderly cat, an idiot bulldog, and a duck. Nothing I thought I couldn't handle.

Plus, I didn't have much else to do at the time. Browsing through life in my late twenties, I was managing a small children's theater and halfheartedly attending art school. My meager paychecks added up to a spare existence based largely on peanut butter and toast, so a three-week house-sitting gig in a fancy neighborhood sounded like just the thing. I packed my bags eagerly, anticipating lazy afternoons poolside.

"Quackers is bereaved," my stepmother announced as she tucked a good navy cardigan into her suitcase for Paris. "He's very lonely. He hasn't been the same since Cheese died."

Cheese, apparently, was the female duck. "He's lonely?" I asked. "How do you know?"

She cringed. "Well, he sometimes has sex with the chickens. But it's fine, really. He's perfectly harmless. Just a lonely widower." She smiled sentimentally. "Would you help me get this suitcase downstairs?"

And that's how I met Quackers, the interspecies duck rapist. Every time I went down to the chicken run, there he was, humping away at a chicken. I couldn't make out the specifics of what was going on, but he'd be up on some poor hen, furiously flapping his wings, and she would look really perturbed. They'd scoot around like this for a minute or two—unless I happened to separate them, which wasn't so easy to do.

"Hey! You . . . duck! Stop it! Stop *doing* that!" Quackers ignored me. I decided it would be useless to appeal to his better nature and instead went for the garden hose. "How do you like *that*, Quackers?" I'd holler, placing my thumb over the aperture to get a high-pressure spray. This usually worked, and both duck and chicken would slink away, feathers soaked and ruffled, looking ashamed.

The garden hose method was effective over the weekend, when I could lounge by the swimming pool and keep an eye on things. But come Monday, I had to go to work, and I scowled at Quackers on my way out the door. "Don't try any funny business, Mister," I warned him. "I'm watching you."

This was a blatant lie, and Quackers must have known it, because when I got back that night the hacienda was eerily silent. "They must have put themselves to bed," I reasoned, taking the flashlight off the shelf by the door and heading down to the chicken run. "I'm sure they're all fine."

They were not. Or at least, one wasn't. A single sweep of the flashlight told me all I needed to know. Feeder, watering can, chicken roost. Three little hens, all in a row. Also a small glass chandelier, because that's how they do farming in Marin County. And one little hen, huddled up on a shelf, crouched in a way that instantly communicated something was wrong. I passed the flashlight beam over her backside, and that's when I saw the blood. I ran for the telephone.

"What do I *do*?" I said in a panic, after calling my then-boyfriend Peter at his job. "It's bloody. It's bleeding. I think Quackers raped it to death!"

"It's a chicken," Peter reasoned. "Cut its head off."

"I can't *do* that!" I squealed. "These chickens aren't regular chickens! They're *pets*!"

And it was true: My dad and his wife had an unnatural attachment to their hens. Besides outfitting their coop in faux Louis XVI décor, Dad and Gail sometimes let them wander around inside the house, where they crapped on the carpets and hassled the bulldog. One time, a framed painting fell down, breaking a chicken's leg. They didn't euthanize it. They took it to the vet and had its leg put in a sling.

Peter's voice was even and calm. "Then you don't have a choice. You'll have to find a vet who's open and take her in."

And I guess you could say that's when my farming career began. I never much cared for animals, unless delicious slices of them were seared in clarified butter and presented to me with a sauce. But now here I was, following my soon-to-be-husband's advice: wrapping the gory chicken in a towel and speeding away to the all-night veterinary clinic. Fumbling at the knobs on my dashboard, I found a Mozart minuet on the public radio station and turned it up, hoping the music would calm her.

Once we got to the clinic, the chicken went downhill fast. "I don't recommend keeping different species in the same enclosure," the vet announced, after I'd been anxiously waiting in his front office for half an hour. "Especially ducks. Ducks are nothing but trouble."

I jumped to my feet. "Where is she? Can I see her?"

Instead of a chicken, the doctor handed me a sheet of paper. It was an invoice with a single item. "Chicken euthanasia," it read. "Cloacal trauma. $345.00."

"Wait—*what*?" I asked, trying hard to be civil. "What's a cloaca? And why does it cost three hundred and forty-five dollars to kill one?"

The vet smiled benignly, murmuring something about after-hours care. And then he told me what a cloaca is.

Unlike human females, who have so many holes we might as well be pasta strainers, the chicken has one perfect, pristine opening, which handles everything. It's her intestinal, urinary, and reproductive aperture. To put it simply, the cloaca is the chicken super-vagina.

I have no idea how they control them. At any moment, this same hole could produce urine, a turd, or a baby chicken egg—a fact that, I imagine, must fill their lives with surprise.

I think it was the knowledge of this elegant organ, much more than the sense of guilt I felt at not being there to protect her, that made me mourn the chicken. The cloaca is so beautifully efficient, such a miracle of avian engineering, that it seemed doubly tragic to think it had been defiled by a sadistic duck.

I got over it, though. My father and stepmother returned from their trip and they tried not to blame me for the death. I went back to my own home, which at that time was a sailboat moored in Richmond, a sketchy part of the Bay Area just east of Marin County.

Slowly, I began to put the horrible incident behind me. But one day, I got to wondering about that duck and why his romantic advances had resulted in tragedy. So I Googled "duck penis," and instantly regretted it.

The duck, I learned, has the longest penis of all vertebrates. When extended, his penis can reach the same size as his full body height, a terrifying ratio when you put it in human terms. The mental image this produced was unspeakable: a sort of Boschian tableau featuring a Satanic duck with a six-foot cock.

Later that week, Gail gave a memorial service for her ill-fated hen. It seems the chicken's name had been Chantal and she'd been as cherished as a miracle baby after a lifetime of infertility. In a troubling blend of fetish and sentiment, Gail had kept the cracked remains of Chantal's first egg over the years, lovingly wrapped in tissue paper. These she placed in a spice jar and buried alongside the bird.

Not surprisingly, my father and stepmother didn't invite me to the wake. This may have been because I was responsible for Chantal's death, or perhaps because my father and Gail had an inkling that I might start giggling at a eulogy for poultry. Whatever the reason, the proceedings went on without me.

To be polite, I did ask about the service, and it seems things didn't go exactly as planned. Gail was vague on the details, but generally, in a case of murder and sexual assault, the assailant is not permitted to free-range at the funeral, pecking at lawn chairs and searching for bugs in the grass. Quackers, however, had roamed the lawn, fixing everyone with a cold, hard stare.

Gail had selected a Shakespearean sonnet to read at the graveside, but no sooner had she begun than Quackers strode lustily forward, causing my father to rear back in alarm. This was a wise move, and he didn't even know the specifics of duck anatomy. A duck penis isn't just huge; it's spiny and shaped like a corkscrew. Clearly, species was no obstacle to this duck's unnatural urges. Who would he fix on next? My father? The bulldog? Chantal's recumbent corpse? The service came to an awkward conclusion, and Quackers was admonished for creeping people out.

Soon after, the bulldog died under mysterious circumstances, and while Gail was convinced he'd snacked on a box of snail poison, I had my doubts. Quackers still roamed the property, terrifying cats and small children, while everyone was careful to keep him away from the hens. He died a year or two later, and though a decade has passed since that terrible night, I still can't feed ducks at the park. "Rapists," I mutter, whenever I see kids tossing bread in the duck pond. "Why feed your *treats* to the *rapists*?"

This has won me some unkind looks from parents and nannies, but they should read up on their anatine anatomy. If they knew what they were feeding, they wouldn't let their kids get so close.

I'd like to say this episode put me off rural living altogether, and for a while, it did. I certainly had no intention of hobby farming, even in the charming way my father and stepmother approached it. But our lives take unexpected turns. Somehow, within a few years, I was managing my own homestead in New Zealand, complete with chickens, goats, and even a few cows grazing in the pasture. The one animal I would not permit on my property was a duck. The very thought of one gave me the creeps.

7 • IT STARTED WITH A CHICKEN

If it seems strange that an artsy San Francisco dilettante should find herself living in a small rural backwater in northern New Zealand, then let me assure you, I'm as surprised as you are. For the most part, our peers back home lead conventionally successful lives: in their early forties, they run businesses, work as lawyers and scientists, have mortgages, and go to restaurants and parties. Meanwhile, Peter and I spend our time chasing cows down the road and executing chickens.

After much thought, I ascribe our unconventional life choices to three main things:

1. The ocean

2. George W. Bush

3. Hobbits

Allow me to explain. For as long as I could remember, sailing for me was a joy. While I was growing up in the San Francisco Bay Area, someone in my family always had a sailboat, and we went out on the bay every few weeks for a dinner picnic, to see the fireworks, or just to spend a day in the sun. San Francisco is known for its brisk ocean breezes, and sometimes howling gusts come shooting down the city hills along the major boulevards. We'd get slammed by a sudden wind, our boat would heel way over and surge ahead, and I'd panic, fumbling for a lifeline.

But my brother Brian was always calm. "Don't worry about it, partner," he'd soothe, patting my stiff yellow lifejacket as he adjusted the tiller. "This boat will not sink." Then he'd explain the physics of sails, how the boat will heel only so much until the sails dump wind, and how she naturally comes back upright, keeping everyone safe on deck. "Always," he said. "She always comes back up."

I loved the wind on my face, the salt on my skin, the pleasant, sunburned drowsiness after a day on the water. Even turkey sandwiches with sour dill pickles tasted better in my brother's cockpit. Most of all, I loved heading out under the Golden Gate Bridge, even if it was just for a day sail. I knew that past that horizon was ocean, then Hawaii, then worlds and continents I'd never seen. It was an endless blue wilderness that I had the power to explore, as long as I kept a strong vessel beneath my feet. And I wanted to get there.

For Peter, the ocean was a sanctuary. Although he's a bright and imaginative man, Peter never did well in the classroom. Now he thinks he probably had an undiagnosed learning disability, but when he was growing up and failing in school, he just felt stupid. On a sailboat, things were different. He had a natural affinity for three-dimensional space, easily understanding which points of sail would chart the most efficient course. He sailed with his father in Long Island Sound, Chesapeake Bay, and down in the Caribbean. And when he was a teenager and his parents separated in an acrimonious divorce, the ocean was the only place he felt calm.

So when we met a decade ago in the laundry room of the Richmond Marina, we were both pursuing the same passion. Peter had quit his job in New York as a network engineer a few months earlier. Instead of marrying his girlfriend and buying a house, he'd poured his down payment into a California boat that would sail him around the world. And I'd just bought *Sereia*, a thirty-six-foot 1970s ketch that I had no idea how to handle alone.

Peter told me his boat was named *Swallow*. "That's so sweet," I replied. "What a cute little bird." Peter's smart-ass friends back in New York had thought a boat named Swallow was racy and hilarious. More than one wanted to know when he'd be getting a dinghy named *Spit*, so I think the fact that I didn't tease him was a relief. We went on a date, we went to bed, and then Peter moved in. Within a year, he sold *Swallow*, and we were planning to sail *Sereia* across the Pacific.

This was 2003 and 2004, the height of the Bush years, and Peter and I were unnerved by the wartime zeal in our country. Most of our fellow liberals were threatening to emigrate to Canada or New Zealand, but as it turns out, we were the crazy ones who did.

It wasn't all high-minded politics that chased us out of the States. The truth was I couldn't afford to live in San Francisco anymore. The city where I grew up, full of artists and poets and revolutionaries, was now home to millionaires. Managing a small children's theater, I had no idea how I'd ever afford to have a family: the cost of health insurance alone for a family of four was more than I earned in a month.

And that's where the hobbits come in. I know it's dumb to say we moved to New Zealand for *Lord of the Rings*, and now that I've lived here for a while, I can definitively say there are no elves. But it's hard to overstate the impact those movies had on silly American dreamers like me—feeling angry and alienated in our own country, just wanting to live somewhere beautiful, where evil wizards are defeated and not voted back into office.

Peter and I started toying with the idea of emigrating to another country, and my first choice was France. "They have national health care," I argued, "and croissants."

But Peter, who had taken first-year French five years in a row, wasn't so sure. "I don't speak the language," he countered. "I don't think I ever will. I won't be able to get a job."

So we did a little research and discovered that most industrialized countries have national health care, including New Zealand. There wouldn't be a language barrier. Peter's IT skills, we soon learned, would put us on a fast track to immigration. And most important, New Zealand was downwind from California. We could get there by sailboat.

Peter moved onto my vessel, and we pooled our resources to throw off the dock lines and head out under the Golden Gate Bridge for an extended sailing voyage. As usual I had no money, but Peter had the proceeds from selling his boat, and a modest inheritance from his father. Figuring that would last us, especially if we caught our own fish and ate like the locals, we traveled through Mexico and Central America. From there we crossed the Gulf of Panama and sailed down to Ecuador and across the Pacific, stopping along the way at pristine tropical islands for cold beer and black pearls. In all those months of travel, I never once thought about chickens, unless I was bartering for a dead one with a local villager.

By the time we settled in New Zealand's North Island, I was six months pregnant. Our first few years were consumed with the challenges of moving to a new country and starting a family all at the same time and not knowing a soul. We wanted to stay on land for a couple of years while our baby was small, and for that we needed a work permit. To get a work permit, we needed a job. And to get a job, we had to move to the coldest part of New Zealand, a place the Rolling Stones once called the "Asshole of the World," a city so cold and bleak they have a hard time finding enough IT professionals, way at the bottom of the South Island: Invercargill.

When our son, Silas, was three months old, we bid our sailboat farewell, packed up the old Mitsubishi we'd bought, and drove eleven hundred miles to the bottom of the country. I took over managing a small youth hostel, and Peter accepted a job at a local technology firm. For fourteen months we toiled down there, bracing ourselves for the freezing Antarctic storms that blew in off the Southern Ocean. When I had a break from the hostel, I'd walk Silas in his baby buggy, a plastic cover pulled tight over his swaddled form as if he were a warm sausage I was saving for lunch. Every now and then, an icy wind would slip under the plastic and Silas would let loose with an ear-rending shriek.

It was in Invercargill that I began to understand the ways New Zealand was different from New York and San Francisco. "Artisan farming" here isn't so much a hipster trend as it is a way of life. There aren't gourmet grocery stores on every block, selling locally sourced radicchio and organic truffle oil. Outside the major cities, most people have a vegetable patch and keep a few chickens. Raising a couple of sheep isn't considered farming, just good sense: you can kill one for the freezer and sell the other for some extra cash.

At first, this was annoying. Why couldn't I get artisanal cheeses at my local Pak'nSave? But a few months into our Invercargill adventure, I discovered the local farmer's market. And that's when everything changed.

We found organic lamb from a farm outside Queenstown, the chops so sweet and tender we called them lollypop chops. We sampled Bluff oysters, some of the best in the world, with the clean, briny taste of the frigid Southern Ocean. We stuffed ourselves with kilos of juicy cherries, and not just one variety of potato, but six or seven, changing with the seasons, each with its own pretty name: Nadines for boiling, Desirees for salads, Red Rascals for a fluffy, creamy mash.

Pushing Silas's buggy among the farmers' stalls, I began to see that without a wealthy population of Wall Street executives and tech entrepreneurs, there just wasn't a market for gourmet shops here. The only way to eat really well in New Zealand was to grow food yourself. Or buy it directly from the growers.

So I bought a few chickens. Hopes held high, I drove out to a local chicken farm with a cardboard box in the backseat. The farmer, a lanky man with oily hair and dirty blue coveralls, opened his barn door to a bedlam of peeping. There were thousands of chicks, tiny golden puffballs pecking at the ground and one another, climbing on their siblings in a roiling sea of cuteness. Without ceremony, he grabbed half a dozen puffballs and tossed them in the box and then handed it to me.

"Thirty dollar," he barked. I could barely hear him over the din.

"That's it?" I asked. "I don't get to pick them out? How do you know they're good ones?"

He looked at me as if I were speaking Inuit. Then his mouth broke into an easy smile, a wide, gummy gap in the front. "All good," he assured me. "Chicken's not good, ya chuck it in the pot!" Then he laughed as if this were a hilarious joke.

After a few months, we had fresh eggs for our family, with a few left over to share with the backpackers. It still didn't occur to me to farm, though. I was too busy running the youth hostel and urging Silas to talk or point or at least say something other than "da." By eighteen months old, he still wasn't walking, either, and we figured he was just a little behind.

But then the tourist season ended, and my job at the youth hostel came to a close. We packed up our things and gave the chickens away to a neighbor. Residency permits in hand, we prepared to move back north, where the sunshine was warm and the winters were mild.

And that's when we confronted the fourth reason for our unusual life choices, the one thing that kept us in New Zealand for good: DNA.

CHAPTER TWO

FREAKY EGGS

At first Silas was nothing but a joy in our lives. He sat up on time, giggled and chortled at all the right moments, and loved being cuddled and kissed. Some things about him were different, but I didn't know they were *wrong*. He never pointed or gestured. He never imitated the sounds I made. And when he cried, there were never any tears. At nineteen months, he still wasn't walking or talking, and soon after, we learned why. Silas has a minute typographical error deep in his genetic code, just a tiny section missing, a flaw so small that it took specialized computers in Australia to find it. It's no one's fault. It's just a fluke.

Silas has a global developmental delay, which means that he's behind in most areas and probably intellectually disabled, though no one knows by how much. At five years old, he could say a few hundred words, but he used only a handful of them, and then only one at a time. "Mih," he would say for *milk*, and "pees" for *please*.

Most of the time I was okay with this, because Silas was also an affectionate little imp. He'd crawl into my bed first thing in the morning and throw his arms around my neck, and when I cuddled him back, his face lit up with joy. The one toy he loved more than anything was a little blue handheld mp3 player, which we called the Dart. I'd loaded this player with recordings of my own voice singing to Silas and reading his favorite books, as well as a selection of Broadway musicals. The music was so calming for him that watching him use the device was like watching a lion get shot with a tranquilizer dart. When his favorite songs came on, when his eyes grew wide with wonder and he hopped up and down with the sheer pleasure of song, I thought to myself that talking wasn't the only way to communicate.

But there was no denying that he made our lives more complicated. It was hard not knowing how Silas would turn out, and it broke my heart to think his life would be limited. There were stacks of paperwork involved in coordinating his care: medical specialists, therapists, teacher aides, and all those appointments for scans, checkups, and tests. And while we took care of Silas's many needs, we still had our savage daughter to tame.

Miranda was born two years after Silas. A typical three-year-old, she chatted nonstop about party dresses and princesses. She also loved riding bikes, jumping on the trampoline, and hacking her own hair off with dull scissors, which meant we usually kept her in a cute little pixie cut.

Our daughter was both sweet and relentless in equal measure. "Mama?" she'd ask, "can I have a juice? Mama, can I have a snack?" Then, once she'd been fed and watered, the real questions began. "Mama, when a crocodile would bite me, would you get a gun and dead him?"

"Of course," I'd say, "of course I would." Then I'd pop another antidepressant, pour a second glass of wine, and dream about moving to the country.

Five years into our life in New Zealand, we finally did it. And Silas was the reason.

Until we became parents, Peter and I both had led the sort of devil-may-care lives where a normal thing, at thirty, was to get on a sailboat and go cruising for a few years. Neither of us was rich, but we were both from comfortable middle-class families whose parents had paid for our educations and thought everything we did was wonderful. Majoring in English and History? "How marvelous!" Backpacking through Central America? "Oh, how rugged!" Working a shitty job in retail to help pay the bills? "What madcap adventures you're collecting for your novel someday."

We'd always been lucky, with the sort of privileged confidence typical of our middle-class, American lives. From the day we met, Peter and I were best friends, compatible in everything, from our taste for good food and travel to sex. We didn't need anyone.

And then we had Silas. When your kid is born disabled, you need. You need help navigating the labyrinth of public services available, from medical treatments to therapies. You need help deciding which interventions to pursue, which might be useful and which will just bankrupt you and disappoint. You need counseling, antidepressants, and wine.

Correction: you probably don't need any of those things. I did. And do. And most of all, what I needed was to know that people in the world would accept my son. For that, we needed a community.

So, I guess, in the end, my brother was wrong. The boat doesn't always bounce back upright. Once we learned what was wrong with Silas, we didn't want to head out to sea. We wanted to come to shore.

When it was time for Silas to start school, I knew he wasn't ready. His language was rudimentary, and he hated holding a pencil. On the other hand, I knew he was intelligent. He had a spark in his eyes, and when he got his first preschool computer game, he picked up the rules very quickly. He might never think like the rest of us, and the signals in his brain misfire, so his language might never be fluent, but I knew he could learn.

Regular school would have eaten him alive. I visited a few of them, watching crowds of first-graders tear nimbly across the pavement, and it was clear my son would have been left in the dust. So I ground up my courage to visit our local special-needs school. The New Zealand education system prefers to mainstream kids with learning difficulties, so the students in special school have only the most serious problems. These are the kids in wheelchairs, the ones you glance at in the grocery store and then quickly look away. I tried to keep an open mind, but when one of the students leaned out of her wheelchair to start chewing on her teacher's skirt, I knew it wasn't the right place for Silas.

Then we found a tiny community west of Whangarei called Purua, in the northern part of the North Island. The immediate draw was the one-room schoolhouse, with just fifteen kids enrolled. It had a vegetable garden, a beehive, and a worm farm. The school was right next to a kiwi bird sanctuary, and when a new egg hatched, the bird keepers called up the school so the kids could come pet the baby kiwi. It was a full New Zealand fantasy, minus the hobbits and elves. When I first visited Purua School, among the rolling green hills of Northland, children greeted me at the door with honey.

Sophia, the principal, looked up. "Will you taste our runny, scrummy honey?" she wanted to know. "The children made it themselves." It was a little freaky. With their vegetable garden, their ceramics, and their ukulele playing, the school seemed weirdly perfect.

Peter and I started calling it "the magical school," not sure if such a place could really exist.

The children were also persistently kind to one another. I brought Silas in for a visit, and he began working with the art program installed on one of the computers. "What an interesting mind he has!" an eight-year-old boy exclaimed. "I didn't even know that program could *do* that." Right then, I knew that if we enrolled our son at Purua, they would take care of him. They would teach him as well as they could, and the children would be kind.

So we moved to the countryside. Peter found work as a network engineer at the local power company, and though he was sorry to let go of our ocean dreams, he was glad for the steady job. And most people were welcoming, despite the fact that I was a goofy San Francisco bohemian sporting sequins and Halloween animal ears in the country. The crazy headgear had started on a whim: I pulled a pair of tiger's ears out of the kids' dress-up basket one day and found they did a great job of keeping my hair out of my face. I added devil's horns and rabbit's ears to the repertoire, and then I just kept wearing them because they made me laugh.

I thought life would be simpler out here. If he wanted to, Silas could eat spiders and run around in the pastures, and if he chose to stand beneath a power pole for half an hour and contemplate its infinite mysteries, no one would judge him for it.

That much was true. But as for simplicity, I was wrong. Instead, our move to the country brought chaos and carnage, and once again it started with chickens. These chickens weren't mine, either. They belonged to Katya and Derek, the nice couple who'd rented their house to us while they spent a year in Germany.

We moved to our new home in February, and within two months, we'd killed all the chickens. This was not our intention, but

19 • FREAKY EGGS

I promise you, it had to be done. The first clue was the eggs. They weren't laying any. From six chickens, I'd get one or two eggs a day, which is a pretty sad return when you're spending thirty bucks a month on chicken feed.

"They're *freeloaders*," I complained to my new country neighbors. "They're *lazy*! It's like I'm running a retirement home for menopausal hens! This has to stop!"

Then there were the freaky eggs. Occasionally I'd get an egg that was malformed, or with a shell that was soft like rice paper. One day I got a deformed chicken abortion: a moist, lumpy mass that appeared to be made from solid shell.

"These chickens are *diseased*!" I announced to anyone who would listen. "There's something seriously *wrong* with them! These eggs are filthy in the eyes of God!"

Maybe it was the filthy God comment, or the fact that I wouldn't shut up, but the school principal, Sophia, agreed to come over and take a look. We decided to combine her impromptu poultry clinic with a dinner party, and I picked up some fresh fish and organic lettuce at our local farmer's market for the occasion.

When Sophia saw the hens, she winced. "Oh, God, they've got *mites*," she said, gasping. "Oh, they can hardly *walk*. They'll *have* to be put down." Sophia is a refugee from a very fancy part of England, and since immigrating to New Zealand, she's lost all traces of her former life except her exquisite taste and her accent. Imagine a willowy woman in a long crimson sundress, with a chilled Pinot Gris in one hand, advocating cold-blooded murder as silver bangles tinkle gently on her sculpted wrist. That's Sophia. She sounded as if she knew what she was talking about. Of course, that's the problem with being American: I'll buy anything if you say it with a fancy British accent.

But it turned out Sophia was right. The chickens had something called scaly leg mites, which are tiny bug parasites that burrow into a chicken's legs and eat the bird from the inside out. They live in tunnels under the skin, just eating and defecating, until the chicken's poor legs are covered with large, scaly sores. Sometimes the chicken's toes fall off. If you let this go on long enough, the hen can't even walk.

So, right before the fish course, Peter went for the axe. "What are you doing?" I asked as he opened the chicken coop door.

"You might want to go back to the house," he informed me.

"You have to *kill* them? For mites? Can't we just . . . do a thing? Put some cream on it?"

"You sweet thing," Sophia assured me. "It's too late for that. The situation has gone on far too long."

I was a little nervous about the ease with which everyone was accepting this Final Chicken Solution, but on the other hand, I had some fresh sea bass in the oven and I wasn't about to overcook it. So maybe I'm a collaborator. All I know is that, twenty minutes later, Peter came back to the house with blood spattered on his trousers and an empty look in his eye.

"No more mites," he grunted as he set off down the hall toward the shower.

I didn't feel great about combining our dinner party with an avian ethnic cleansing, but I did understand it was necessary. There comes a point when an animal is so sick, or so crippled, or in so much pain, that the real cruelty is *not* euthanizing it. The day after the leg mite incident, I took great care to make sure it wouldn't happen again: I pulled on my rattiest clothes and grabbed a shovel and a bottle of bleach. For an entire day, I scrubbed that henhouse till I felt confident there wasn't a leg mite in sight. I shoveled shit, scrubbed the walls, scraped the floors, and sprayed the whole thing down with farm disinfectant. The next batch of chickens, I vowed, would be sheltered from mites and rapists. The next batch of chickens would be safe.

And for a while, they were. Until they got leprosy.

This was not as terrifying as it might seem. Five years previously, I'd witnessed the same thing with my Invercargill birds. Back then, my first batch of chicks lived in a big wooden crate in our living room. I clipped a lightbulb to the box to keep them warm, and I changed their wood shavings daily. But then they developed beak rot. Something was terribly wrong.

I jumped on Google to investigate. "My chickens have leprosy," I posted on a variety of chicken forums. "Please help." And people did, from all over the world. There was a chicken enthusiast in Tennessee who suggested they were pecking one another, and some guy in Adelaide who thought they might be banging their beaks on the metal bars of their cage. "No cage," I insisted, increasingly alarmed. "These chickens have never even seen metal. What do I do?!"

After several days of frantic messaging, the chicken community was stumped. And then the beaks started falling off. Not the whole beak, exactly, just the tip, like in the photos you might have seen of sad people in leprosy camps.

Hysterical, I called the Invercargill chicken farm. "What's wrong with these chickens?" I demanded, trying to sound a little bit sane. "*Their beaks are falling off.*"

There was a pause on the line, while the farmer tried to place who I was. "Aw, the beaks?" he asked finally. "Aw, yep. Those have been lasered."

"*Lasered?*" I repeated. "Why would you put a *laser beam* on a chicken beak?"

"Saves them from pecking each other," he explained, as though this were obvious. "Doesn't hurt them. Just kills the tissue. After a few weeks, the tip comes off and the beak is nice and round."

"Oh," I replied, ashamed. I had learned a lot of things growing up in San Francisco, such as how to speak French and prepare for a major earthquake. I had not, however, learned about laser beams and chicken beaks. I thanked the farmer and hung up the phone.

So now my second batch of chickens had beak rot, and I was taking it like a pro. I felt rather motherly toward my hens, so it wouldn't have been so easy to execute them for minor infractions such as leg mites. I fussed over them daily, making sure they could still peck for bugs with their artificially stunted beaks. I studied their behavior and announced my discoveries at the dinner table.

"They wallow in dirt," I reported, once the chickens were old enough to live outside. I hadn't noticed this phenomenon with my first batch of hens.

Peter was skeptical. "Wallow? In dirt? I thought *pigs* wallowed."

"No, it was definitely dirt. They were rolling it all over themselves. It was filthy."

Silas sat beside me not eating his dinner. Hearing a word he knew, he joined the conversation. "DUHT!" he yelled. "Duh-tee!"

"I'm dirty, Mama," Miranda joined in. "My *butt* is dirty!"

"Don't use potty talk," I snapped. To Peter, I asked, "Do you think they're crazy? Maybe we should Google it."

A little poking around on the Internet revealed that dirt wallowing is actually normal chicken behavior, called a dust bath. They roll around in the dirt, cover themselves in filth, and this rinses out the mites. So dirty chickens weren't our problem.

Our problem was the cow.

When Katya and Derek agreed to rent their home to us, they

gave us a big discount off the cost of their mortgage because the house also came with a number of responsibilities. There were six chickens to care for, an old shaggy dog named Phoenix, a couple of cats, and a large, probably pregnant cow.

This cow was supposedly named Lucky. I did not name her. If it had been up to me, she'd have been christened A Thousand Pounds of Anarchy, which would have been a lot more descriptive and, frankly, more accurate. Because Lucky jumped fences.

I'd been wary of this cow from the start. A hefty black Friesian with a ghostly white face, Lucky looked slightly malevolent to me, like an evil sorcerer engaged in some violent, black-hearted rite. Also, I had no idea how to care for her.

"So what does she need?" I'd asked Derek before he left, hopeful that he'd grace me with cow pointers.

"Aw, nothing, really," Derek assured me, patting her affectionately on the snout. "She'll eat the grass and stay in her paddock. Just make sure she has water to drink, and she'll take care of herself, eh."

Which was mostly true, apart from "stay in her paddock" and "nothing, really." Actually, none of it was true. Because Lucky was trouble from the start.

To begin with, she didn't like grass. She'd much rather have eaten the tender new pea shoots or the sweetest baby lettuces from the garden. As soon as our backs were turned, she'd jump the fence. Then we'd find her in the veggie patch munching on organic chard or out on the road taking a pleasant stroll toward town.

As well as Lucky and the rest of the farm, we had our own family pets: a black-and-white cat named Catty and a hyperactive German shepherd puppy named Kowhai. Kowhai (pronounced KO-fai) is the name of a delicate New Zealand tree with tiny leaves and clusters of yellow blooms. It's a very feminine tree, and a terrible name for our dog, who should probably have been named Mayhem or possibly Fang. She wants only to play, but unfortunately she weighs eighty pounds and has bone-crunching jaws, so sometimes people misunderstand her intentions.

You might say that things were not going swimmingly. One month into our stay, we'd managed to dispatch most of our charges: We executed the chickens. One of the cats disappeared, clearly disgusted with our urban ways. And Lucky was escaping almost daily. It seemed we didn't have much of a talent for farming.

And we still had eleven months to go.

CHAPTER THREE

PARTY TIME

Most of Purua is split up between five or six original families, the descendants of pioneers who established big farms on enormous landholdings. These old-school farmers had their own way of doing things, and they weren't really interested in hearing outside opinions, especially from an American woman wearing devil's horns. The first time some of our new neighbors invited us to dinner, we disgraced ourselves immediately.

Shortly before Silas's first day of school, a couple named Karl and Catherine invited us to their home—and the first thing we did was pee on their rug. Not Peter or me, but Kowhai, who was just a puppy at the time. "May I bring our dog inside?" I'd asked Catherine, thinking this was a casual country affair and that everyone liked hanging out with puppies.

"Er, sure," she stammered, which I took to mean yes, but which actually meant "That's the most disgusting thing I've ever heard, because aren't you aware that dogs eat their own shit?" The evening deteriorated from there. Silas started chasing their chickens, which seemed like a normal thing to do, except apparently there were lots of baby chicks in the flock, and some of them got trampled in the melee.

"Bud! Bud! Bud!" Silas crowed, tearing after the hens with a huge grin on his face. The chickens scattered in hysterical panic, leaving a few broken fluff balls in the dust.

"Oh, my God. I'm so sorry!" Peter stammered when he saw what had happened. He reached for a chick. "Are they going to be okay?"

"Probably not," Catherine murmured. "They're dead." Then she tried to make us feel better. "But these things happen sometimes on a farm."

Given the fact that we'd defiled their home and started a poultry stampede, I wasn't surprised that our friendship with Karl and Catherine didn't bloom right away. As Karl liked to say, "You're not really in this community till you've been here seven years," which seemed like a high bar to set for the rest of us. But since his family had been there for about a century and a half, I could see where he was coming from.

To our relief, serious professional farmers weren't the only ones who lived out here. There were also the "lifestylers," and that's the group we fit into. "Lifestyle farmers," "hobby farmers," or, for the pretentious foodies, "artisan farmers"—we're all pretty much the same. We're the pretend farmers, the ones who live on a few acres of land, keep some animals, and for the most part maintain professional jobs in town. We raise chickens, heirloom tomatoes, and cats, while the real farmers raise cows and sheep, for milk or meat.

The first lifestyler couple we befriended was Nick and Amanda, because Nick was an occupational therapist who'd worked with Silas for a number of years in town. A Malaysian guy who'd grown up in New Zealand, Nick was the one who'd patiently taught Silas how to use a spoon and pedal a trike. When Silas was driving us nuts with his moaning, so much so that we felt we were being haunted by a midget ghoul, it was Nick who calmly suggested we make a "log of the moan," describing when Silas moaned and what was happening at the time. Very quickly, we came to realize that those awful sounds weren't random. Silas would moan when he needed something, such as water or food, but didn't have the words to ask. Nick was so gentle and intuitive with Silas that we were surprised to learn he spent his spare time as a committed martial artist. The guy was not much taller than I, and he could kill most people with his bare hands.

There were other surprises. Despite his professional decorum, Nick would often laugh out loud at my off-color jokes—such as when I described my disabled son as a "midget ghoul." I was delighted to find that his wife, Amanda, was just as irreverent, and sharp as a whip.

Five feet nothing, with thick dark hair and fierce black eyebrows, Amanda was also an impressive mother. Her three daughters were intelligent, confident, and beautiful, with high cheekbones and wide, dark eyes. There was Sophie, a thoughtful and serious nine-year-old; Amelia, who at five was all sweetness and cuddles; and Lucy, the baby.

In our first weeks in Purua, I closely observed Amanda's parenting style, which seemed to consist of ignoring the children while they ran around outside and tackled sheep. At first I expressed some concern: Wouldn't they get lost in the forest? Fall into a river? Get mauled by a possum?

"This is your life now," Amanda informed me, sipping a chilled Sauvignon Blanc. "You just let them run off, then sit down and have a wine. They take care of each other."

She seemed very relaxed about it, so I tried to adjust. *This is how you raise kids in the country*, I considered. *You lounge around and drink, while they tear through the bush like a pack of wild dogs.* This suited me just fine. Actually, it was everything I'd ever dreamed of: the moral high ground of parenting without any of the effort.

Shortly after we moved to Purua, Amanda and Nick invited us to Lucy's second birthday party. It was how I'd imagined country living, with red-and-white-checked tablecloths and a birthday cake shaped like a teddy bear. There were chickens scampering in the grass and a beautiful pearly white goat with a slender black stripe down the middle of her back. Her ears were long and silky, framing her face in a feminine way. When I saw her, I caught my breath.

"I *love* your goat," I told Amanda, who was rushing around with a tray of fruit salad.

"What, Pearl?" Amanda asked, setting paper plates in front of the children. "You can have her. Come and get her anytime."

"You don't want her?"

Amanda cocked her head and looked skeptically at the goat. "It's not that . . ." she hedged. "Pearl's lovely. But goats are a lot of work." She turned her eyebrows on me. "A *lot* of work."

I should have listened, but once Amanda told me I could have her, Pearl was all I could think about. I sat down with the other mothers and helped myself to a plate of fruit salad. *We can get milk from a goat*, I fantasized. *We can make cheese!*

The kids raced around solving riddles for a scavenger hunt, while the mothers languished idly in the shade. Abi was one of the first women I spoke to at the party. With wide blue eyes and blonde hair in perpetual disarray, Abi gave the impression she was constantly overwhelmed. She claimed this was because she was frightened of her children, but I had my doubts. Since she worked as an occupational therapist while raising two kids, she probably had her life in order.

"Does lettuce give chickens diarrhea?" Abi wanted to know. "Because I read it gives them diarrhea, and now I'm not sure what to do with my old lettuce anymore."

"Vinegar," suggested Michiko. Michiko seemed like a very sweet lady, but I didn't know much about her except that she worked as an accountant in town and was an accomplished classical pianist. Her Japanese accent made it hard for me to understand her. "Feed vinegar," she repeated. "And cinnamon. Very good for chicken."

We discussed the finer points of chicken bowels, and then we moved on to husbands.

"I heard Nick's really into martial arts," I ventured. "Which one does he do?"

Amanda rolled her eyes. "Which one *doesn't* he do, more like. Karate, ninjutsu, kickboxing. Now he's into this thing called Krav Maga. They developed it for the Israeli military so unarmed soldiers could kill two people at once."

Abi's blue eyes widened. "Oh, my gosh," she said. "It's all I can do to get Zane to mow the lawn."

Autumn leaned forward, reaching for the bowl of fruit salad. She struck me as an interesting woman. With curly brown hair, broad shoulders, and formidable curves, she nonetheless had a delicate way about her, but she didn't mind being brutally honest. Also, she loved food. Her husband, Patrice, was a French chef, and she'd rather have talked about food than martial arts.

"Kids went eeling the other day," she volunteered. "Caught a bloody big eel, and now Patrice has to take its skin off with pliers."

"You eat *eel*?" I asked.

"Ah, it's *gorgeous*," Autumn told me. "S'long as you scrape off the slime coat."

"Slime *what*?"

"You boil it in a pot, you see," Autumn explained. "Then the slime goes white, like an egg white. You scrape it off, and it's fabulous."

I leaned back in my chair, taking in the view of rolling green hills dotted here and there with little sheep frolicking in the sunshine. And that's when the screaming started.

"Mama!" Amelia howled. "*My eye! Miranda stabbed me in my eye!*"

I glared at my three-year-old daughter, who had somehow got ahold of a vegetable peeler. She had a peculiar smile on her face that I'd seen only in horror movies. "*That's it!*" I shrieked. "*You're going to sit in the car!*"

Amanda reached for Amelia, examining the wound. "Her eye's fine," she noted. "She's not blind."

Silas had been quiet for some time, but just then I heard him yelling from the far side of the house. "*Toit!*" he hollered. "*Toit! Toit!*"

Peter broke off from the cluster of local dads he'd been chatting with. "What the hell's he saying?" he asked.

"Beats the hell out of me," I said, shrugging. "He's an alien." I was wrestling Miranda into the car, where I planned to incarcerate her until she calmed down. My strategy was failing. Far from calm, Miranda was now hurtling herself against the glass like a tiny meth addict.

"Are you ready to come out?" I called.

"Yes," came a meek voice. I opened the car door. "Mama, what the hell?" Miranda scowled at me. "What the hell is you doing?"

A mind-bending shriek tore through the air, the kind that makes you think of axe murderers and their hapless victims. We rushed to the far side of the house expecting blood, only to find Silas bent over in the grass giving birth to a well-formed poo. For some reason, this always upsets him.

"You okay, Silas?" Peter asked, approaching. "Do you need help?"

Silas picked up the poo and admired it, then presented the gift to his father. "TOIT," he announced.

"Toilet," I suddenly realized. "He wants to put it in the toilet!"

"Yes, it should go in the toilet," Peter corrected, "but you're outside now. Just throw it in the bushes."

Silas was having none of this. He clutched the poo like a cupcake in a world full of famine.

"Drop it," Peter ordered.

This time the poo tumbled to the ground. A look of consternation passed across Silas's face. He bent down and picked it up, then brushed off some stray bits of dirt from the turd. *"Toit,"* he persisted.

Peter threw up his hands. *"Fine.* We'll take it to the toilet."

Attempting a casual smile, I glanced back at the party guests. Amanda was still clutching her wounded five-year-old, and the others were openly staring.

"It's fine," I assured them. "We're fine. Just . . . kids being silly!" I went back to sit down with them, casually popping a grape in my mouth.

Amanda put her hand over mine. "You might not want to call

him an alien. You know, in front of people. They might get the wrong idea."

I smiled and agreed, because I didn't know how to explain that it's easier to think of Silas as a perfect little alien with different social norms than a developmentally delayed human child who wreaks havoc at parties.

Maria, who is from England, seemed expert at changing the subject in an awkward situation. "Did you want any beef?" she interjected. "We're slaughtering a steer this weekend. It's six hundred pounds of meat, so we're looking for someone to go halves with."

This gave me pause. I turned to her, noting her long, shapely legs and firm upper arms. She didn't look like a person who handled beef by the hundreds of pounds. "*Three hundred pounds?*" I repeated. "That's a lot of cow."

"Well, it's for the whole year," she explained. "You just chuck it in the deep freeze, then you eat it the rest of the year."

I thought about this. "But isn't that, like, six pounds of beef every week?"

"Yeah, but it's great meat. Grass-fed! You'll love it."

I wasn't sure how I could backpedal out of buying half a steer's worth of heart attack, but in the end, I didn't have to. Because that's when Kowhai bit the baby.

"*Ow!*" Lucy screamed. "*Mama!*" All the mothers instinctively scrambled to their feet, and there was Amanda's husband, Nick, dangling Lucy over Kowhai's head. She looked like bait.

"*Kowhai!*" I roared. "*No biting!*"

Kowhai ran to me looking guilty, her tail between her legs.

"It's fine," Amanda assured me. "Don't worry about it. Lucy's fine. She's just a little drama queen."

I cast a wary glance at Nick, who had his eyes locked on Kowhai.

With his arsenal of international killing skills, Nick could probably have dispatched our dog with a flick of his wrist. At this point, we'd stabbed one child in the eye, shat on his lawn, and savaged the birthday girl with our German shepherd. If he chose to go Krav on us, we'd have had only ourselves to blame.

"You're sweet," I told Amanda. "But I think it's time to go home."

CHAPTER FOUR

TEDDY BEAR CAMELS

A few weeks later, Peter went for a hike in a local nature preserve, leaving me home with the kids. And that's when the cow got out again. I checked on her in the morning, when I went out to feed the dogs and the chickens. She was just standing there in her paddock, looking innocent and chewing her cud. But then I took the kids to the grocery store. And when we got back, she was nowhere to be found.

Now, if your chickens get out, it makes a charming rural tableau. And if your dog gets out, it's a mild annoyance. But when your cow is staggering down the highway snacking on daisies and dodging sedans, it's a big fucking problem. If the cow causes an accident, you're legally liable, not to mention the possible carnage.

Most of the time, we got her back in her paddock. Peter and I would go running into the garden with long sticks of bamboo, yelling "*Baaaaa!*" and "*Go, cow, go!*" which are technical terms for cow herding. Most of the time, this worked. But sometimes Lucky would disappear for days, and we could only hope she was staying away from moving vehicles.

As far as I was concerned, she could stay gone, but we had promised Katya and Derek we'd keep her safe. And there was the unpleasant chance she could be hit by a car, which would have been messy, and probably dangerous for the driver. So I had to retrieve her. I wasn't all that sure how to do this by myself, as cow herding wasn't a subject they'd covered in liberal arts school, but I thought I'd improvise.

"Stay here," I told Miranda. *"And don't get out of the car."*

"Okay, Mama," she said, nodding. "I promise. You go get Lucky?"

"Yeah," I told her. "Something like that."

Silas, who was listening to his Dart, said nothing.

So I grabbed a stick—because even if I had a lasso, I wouldn't know what to do with it if it slapped me in the ass—and went off to find the cow.

Lucky was easy to locate once I got to the road, standing as she was in a patch of kikuyu grass and lowing at passing cars. I endeavored to catch her attention.

"Baaa!" I screamed. *"Baaa!"*

Despite my efforts, the cow did not interpret this clamor as a request to return to her paddock. Instead, she correctly assessed me as an amateur without a clue. Ignoring me utterly, she ducked into our neighbor Hamish's garden.

This worked out fine, because Hamish is a dairy farmer. So while he probably would have preferred not to have our cow gobbling his roses, at least he knew what to do with a four-legged beast when he saw one. Besides pick up a stick and yell, "Baaa!"

Then the real problems started. "Mama!" I heard. "I'm cooooooming!"

That sounds a lot like Miranda, I thought, with impressive naïveté. *But it can't be Miranda, because she promised she wouldn't get out of the car.*

That's when I noticed my three-year-old trotting down the center of the highway. Her height was just right for this, because she was short enough to be completely invisible to any speeding motorist.

"*Miranda!*" I hollered in a terrifying roar I did not know I possessed. "*Get. Out. Of. The. Road. Now!*"

Predictably, she burst into tears. But she got out of the road.

"I'm sorry, Mama. I'm sorry," she sobbed. "*Please!* I want to be your *friend!*"

So I slapped her.

No, just kidding. I snuggled her up and held her close and carried her back to our property, where Silas was still in the car. Like a little angel child, he had done what he was told.

I opened the car door.

My son was squatting in the passenger seat clutching two patties of warm poo. Having already spread a good deal of it over the car's upholstery, my purse, and my cell phone, he seemed unsure what to decorate next. He looked to me as if for inspiration.

I took a deep breath and counted to ten, just as the anger management people suggest. Then I plucked the patties from Silas's hands and flung them in the bushes.

With two delinquent poo incidents so close together, I don't want to give you the impression that Silas is one of those tricky poo-obsessed children who smear it on the walls or flick it around. Now, at five years old, he was basically toilet trained. The problems arose when he wasn't near a familiar toilet and didn't have the words to ask for one. Had he been able to speak, I feel confident he would have asked to be let into the house so he could use the bathroom like a normal child. But he didn't know how to ask. So he had had an accident. Then, faced with a number of rogue turds in the back of his mother's car, he had done the only other thing he could think to do: he spread them on things.

I made sure Hamish's front gate was closed. I knew Lucky would be safe across the road for a few days, but as far as I was concerned, that solo cow-wrangling escapade was the final straw. Country life was complicated enough without dodging traffic to chase after livestock. We asked around the neighborhood and eventually found a farmer who was willing to let Lucky eat his grass until Katya and Derek came back. He came over to pick her up, and A Thousand Pounds of Anarchy left our lives forever.

Now we just had the young chickens to care for, in addition to a few dogs and cats. It seemed time to expand our menagerie. If that comes as a surprise, considering it was April and we'd already banished or murdered the majority of the animals we'd had charge of, then I must tell you one thing: Peter hates computers. He knows it's the best way he can earn a living, and he feels fortunate to be able to support his family, but basically he hates working at a desk all day. He's always casting about for business ideas, preferably ones that involve his working outside.

"I'm thinking alpacas," he announced one day, tapping away at his tablet. It was Sunday morning and we were sipping our coffee in peace. Phoenix, the shaggy old dog, sighed contentedly on the floor. Silas and Miranda were still snoring soundly in the other room.

"Alpacas?" I asked. "Isn't that like a llama?"

Rolling his eyes at my ignorance, Peter showed me his tablet. "Alpacas are a *type* of llama, but not all llamas are alpacas. See?"

The Google page he had up displayed a procession of alpacas, each one cuter than the last. If a teddy bear made it with a camel, that's what their babies would look like. Alpacas are long-necked, soft, and fluffy, with enormous dark eyes and the most insanely long eyelashes I've ever seen. They come in many colors, too: white, black, and spotted, and a color called rose-gray, which I swear to God looks like purple. That's right: purple teddy bear camels. That's how cute these things are.

"How much?" I asked, breathless with desire.

"Twenty thousand dollars," Peter said. "For a good one."

I swallowed abruptly, and hot coffee flew up my nose. "Twenty—*what*?! For a *camel*? We're saving that money for a *house!*"

"Not camels. *Camelids*," Peter corrected me. "And I'm thinking it's a good investment. We could produce fine fiber, even sell it later on. Then, when we get a second one, we can breed them and sell crias."

"So *forty* thousand dollars," I repeated dumbly. I reflected on this use for our life's savings. "And sell crack?"

"No, *crias*," Peter repeated. "Baby alpacas. They're very cute." He typed into his tablet and showed me more pictures, this time of *miniature* purple teddy bear camels.

"Jesus," I commented. "That's fucked-up."

"I know, right?" Peter finished his coffee, a big grin on his face. "It's a fantastic business idea."

Later that day, I ran this fantastic business idea past Autumn, who is an actual New Zealander who grew up on a farm. Unlike either of us, she also had some experience working with fibers. I sat at her long wooden dining table while she bustled about the kitchen making coffee. She did this with surprising dexterity, since she was balancing her three-year-old son, Titou, on one hip while juggling sugar and hot water.

I noticed that Autumn didn't proffer her opinion lightly. Instead, she collected the data. And she did this by peppering me with questions.

"So, what's the product you're trying to sell? Would it be fleece or—"

"Pashminas," I said. I don't know why I said pashminas. I think I just wanted one.

"So, very large scarves. And you'd sell these for—"

"Five hundred dollars."

Autumn sucked her teeth. "You wouldn't get much of a market for five-hundred-dollar scarves here in New Zealand."

"No, like, to wealthy tourists. And maybe we'd ship overseas."

Maris, Autumn's younger daughter, slipped into the room. She was bleeding from a jagged head wound. Ignoring us, she reached for a well-worn paperback and curled up on the soft wicker sofa.

"Uh, Maris?" I asked. "Are you okay?"

She smiled benignly, causing the lumpy red gore to dribble down her temple and onto her cheek. Her elder sister, Nova, wandered in and pulled up her sleeve to reveal a slashed wrist. She sat down in the armchair and got out her sketchbook, at which point I noticed twin incisions at the base of her neck.

"Autumn?" I raised one eyebrow. "Is there something wrong with the girls?"

Autumn rolled her eyes. "Would you take it outside, please? I'm tired of cleaning tomato sauce off the furniture."

"She dead now?" Titou wanted to know.

"Yes." Maris glanced up. "I can't go outside. I'm dead."

"Maris," Autumn snapped. "*Now.*" She stepped toward her daughters, and the two girls screeched and ran, dropping their books as they fled.

Autumn shook her head, gently lowering Titou to the floor. She poured the coffee into mismatched mugs, one of them hand-painted with hearts and polka dots by a child. Then she sat down across from me. Titou crawled beneath the table, where he settled in with a coloring book and some crayons.

"Sorry about that. I don't know if it's the *Twilight* books or what. They've just got so *morbid* lately. Maybe we should get a television." Autumn sipped her coffee pensively.

"Please don't," I urged her. "I think they're wonderful."

"We do go through a lot of tomato sauce, though. And they scare their brother." She checked under the table. "You all right, Titou?"

Titou, engaged in covering a page with a crimson crayon, said nothing.

"He's fine. But anyway. About your business." She put down her cup and clasped her hands in thought. "Look, I just don't see it. You'd have to card the wool, then spin it and weave it. You'd need a whole fleece for a scarf that size. So if you're doing it by hand, then carding—that's five hours; then ten hours of spinning and maybe another twelve of weaving. Don't forget setting up your warp; that's another few hours."

"So, like, a week?" I asked.

Autumn nodded.

"For a scarf?"

She smiled. "And minimum wage in New Zealand's coming up on thirteen dollars an hour."

I did some quick math in my head. "So basically it would cost more to make the scarf than I could ever sell it for."

She nodded.

"Then there's equipment and taxes and shipping," I said, ticking the items off on one hand.

Autumn kept nodding, a smile playing about her lips. She was beginning to enjoy this.

"And I'd still have to pay for the twenty-thousand-dollar camels."

"Who'd produce only one fleece a year."

I set down my coffee cup, shut my eyes, and massaged my temples with my fingers. "Why did I marry an English major?"

Autumn laughed. "Don't ask me! I married a French chef, and he's retired. We're always broke, but we do eat well. D'you want some pineapple upside-down cake?"

"Yes, please!" came a voice from beneath the table.

After stress-eating cake at Autumn's, I went home to Peter and suggested that possibly twenty-thousand-dollar alpacas were not the greatest business idea ever.

"You're always crushing my dreams," he grumbled. "Dream crusher."

"Tell you what," I suggested. "Let's just get a couple of cheap ones. Discount alpacas. Let's see if we even *like* them before you start your camel empire."

So we started calling local breeders. And that's how we met Gay and Mike, alpaca enthusiasts who are also insane. A former stay-at-home mother and a computer programmer, respectively, these two had retired from their sensible lives to breed alpacas. They purchased show-quality animals, sometimes for tens of thousands of dollars, then sold their babies to neophytes like us.

"Oh, they make wonderful pets," Gay crooned to me over the phone. And she's right. Alpacas are a perfect livestock for the beginning farmer, because they're very expensive and totally useless. Bred by the ancient Incas for their fleece, they produce no meat, milk, or eggs. Instead, they give off attitude. Lots. And lots. Of attitude.

But at the time, we didn't know that. I just thought they were crazy cute.

"When can we come round to see them?" I asked. We set a date, and the following weekend, Peter and I found ourselves driving out to a real live alpaca farm. When we parked in the driveway, everything seemed relatively normal. Gay and Mike were a couple in their late sixties, both with white hair and skin tanned gold from their work outside.

"So glad you've come to meet the boys!" Gay greeted us with a wide smile. "Come, come inside."

Miranda tugged on my hand. "Mama, are those 'pacas?"

I looked where she was pointing, and indeed they were. Dozens of alpacas in every imaginable shade: black ones, brown ones, white ones, and purple ones. Their necks were long and elegant, and they had little puffballs of fleece at the tops of their heads. One of them even had polka dots.

I admired him, but Mike coughed politely. "That's Rabbie," he told me. "He's not for sale."

I turned back to Peter and Silas, who were just getting out of the car. "Silas," I coaxed. "See the 'pacas? Aren't they *cute*?"

Silas took one look at the beasts and scrambled back in the station wagon. "No! *No!* Home!"

Mike looked concerned. "Is he all right?"

"Fine," Peter assured him. "He just has trouble with transitions."

But as it turned out, Silas was the sensible one. Beneath their cute façades, alpacas are dangerous beasts. We should have sped back to the safety of home, but instead we went inside for a cup of tea. Into the cold-blooded heart of alpacadom.

Displayed in what was otherwise a conventional living room were hundreds of alpaca sculptures. *Thousands* of them. There were dolls made out of alpaca fleece. Alpacas made of alpaca fleece. Inexplicably, there were decapitated alpaca heads arranged in rows along the wall.

"They're not *actual* heads," Gay informed me. "I didn't sever them. I felt the fleece, then I sculpt with it." She giggled. "I guess I'm a little obsessed."

Arrayed on the mantel was an assortment of alpaca figurines, and above them hung Gay's masterpiece: an enormous tableau from *The Wizard of Oz* sculpted entirely from alpaca fleece. I gaped.

"Took me weeks," Gay called from the kitchen. "Sculpting Dorothy and Toto, then the Scarecrow and the Lion. Tin Man was the hardest bit." She emerged balancing a tray full of tea things. "Tried to send it to my grandson off in London, but the shipping was too dear."

I looked at the tapestry. It was a full square meter of alpaca fleece, everything rendered in bas-relief: the characters, the yellow brick road, and a multicolor rainbow arching over the Emerald City. That poor English grandson; he'd really missed out.

"The thing is, they're rather addictive," Gay explained while sipping her tea. "Once you've had alpacas, you'll never *not* have alpacas, if you know what I mean."

This should have come as a warning. Features in your life that you can't get rid of—a heroin addiction, for example, or genital herpes—are not necessarily good things. But by that stage, I was too far gone.

I set down my tea. "Let's go meet the boys."

Mike had corralled the older male alpacas, the ones priced between three and five hundred dollars, into a pen near the house, so we wouldn't be tempted by the pricier ones. Silas was calm now, so Peter held him up near the pen and pointed out all their cute features.

"See how fluffy?" Peter asked.

"Tuh-tee!" Silas repeated.

Gay came up behind me. "You know, alpacas cure autism," she offered. "Really. There are some amazing stories out there. I'll have to send you some."

Peter and I exchanged a look. We've long since stopped looking for ways to "cure" Silas, and instead we just love him for the weird kid he is. Also, he's delayed, not autistic. But Gay was just trying to help, and the teddy bear camels couldn't be *bad* for Silas. Or so we thought.

"Can we get all of them?" Miranda wanted to know. "I want *all* the 'pacas."

"No way," I reined her in. "We're getting only two."

Mike, who had been quietly scratching them behind the ears, coughed discreetly. "Oh, you can't get just two," he informed us.

"Why not? Because they're addictive?"

"No," he replied, quite seriously. "These are herd animals. If you get just two, and then one dies, the other will die of a broken heart."

This had to be some kind of a sales ploy. "Are you serious?" Peter asked. "A broken heart?"

"No, Mike's right," Gay assured us. "You'll have to get at least three. They get so attached to each other, you see."

Then she relayed the alpaca miracles. There were, it seemed, alpacas who hummed to autistic children, making them stop rocking and hitting themselves. There were alpacas who let old people pet them, and then those ancient cripples would rise from their wheelchairs with renewed vigor. I wouldn't be surprised if alpacas had cured cancer and caused the blind to see. "They really are magical animals," she effused.

What she didn't tell us is how much distance they get.

We heard a quick rush of air, like someone popping a beer bottle. "Oh, Kenny." Gay sighed, wiping the front of her T-shirt with a rag. She was covered in green slime, and it smelled like an open grave.

"Mama, did you *fart*?" Miranda wanted to know. "It smells like *fart*."

"Oh, it's just a little saliva." Gay smiled sheepishly. "Did I mention they spit?"

"Nope," Peter said. "You didn't mention that."

"It's very rare," Mike explained, shifting Kenny to the back of the pack. "They don't like the smell of it, either. See how his lower lip is hanging? That's to get rid of the smell."

"Oh, I hope you'll still adopt them," Gay urged. "I promise, you won't regret it."

So, because we are stupid, we bought three. And they were beautiful. Kenny, Henri, and McTavish were each a different shade of elegant: white, tan, and a deep, chocolate brown. "That's my pashmina," I informed Peter, pointing at McTavish. "At the very least, I've got to make one."

At first, things went pretty smoothly. Mike and Gay sent us home with a pack of alpaca information: we had each animal's lineage, and descriptions of the various ills that could befall them, such as facial eczema, ingrown toenails, and something called rye grass staggers.

But there were other things they didn't mention. Such as the fact that alpacas have "fighting teeth." These are long, razor-sharp canines that they sequester in the back of their mouths like daggers. Our teddy bear camels had fangs.

And when they started working out their place in the hierarchy, our alpacas turned evil.

Gay and Mike had warned us this would happen, but as usual

they'd played it down. "Might just take them a few days to get used to each other," Mike mentioned as he backed their truck down our driveway. "Shouldn't be too much of a bother. They'll need to work out who's boss."

Then he and Gay smiled and waved and left us alone with the alpacas.

We filled three purple bowls with alpaca nuts and set off for the paddock.

"Can I bring my That Baby to the 'paca friends?" Miranda wanted to know. "That Baby" was the first doll we'd ever given her, back when she was eighteen months old and just learning to talk. She'd called it 'Dat Baby, and the label had stuck.

"Of course you can bring her," I allowed. "Come on! Let's go feed the boys."

"Silas, you coming?" Peter called, but Silas shook his head. "*No, no,*" he insisted, hugging his Dart to his ear. He was listening to *Peter Pan*, and the grin on his face was huge.

"Have fun," Peter said, shrugging, and the three of us set off, Kowhai trotting along behind. Phoenix, who was older and more sensible, opted to continue sleeping on the front deck.

We should have stayed home with Peter Pan and Phoenix, because when we got to the paddock, the alpacas had changed.

Kenny, Henri, and McTavish stood there dead-eyed, lower lips hanging slack. Their mouths gaped, exposing rows of yellowing fangs. There was a viscous green fluid collecting on their tongues, sticky streams of it spilling onto the ground.

And then they started to moan. "*Beeeeeeehn,*" they groaned. "*Beeeeeeeehn.*"

"Did they just say, 'Brain'?" Peter looked at me. "I think they said, 'Brain.'"

The alpacas were advancing, green drool pooling at their feet. They were very close now, no more than five or six feet away.

"*Beeeeeeeeehn* . . ."

"Gay says it's humming," I said brightly. "But it sure sounds like 'brain' to me."

"*Beeeeeeeeehn* . . ."

"Mama?" Miranda reached for my hand. "What's wrong with the 'pacas?"

"Well, Magnolia," I explained, using my special pet name for her. "I think they want to eat our brains."

"Look at Kowhai," Peter whispered. Our German shepherd was crouched low like a wolf. Hackles raised, she crept in for the kill.

Then the trance snapped. Kenny spat and charged, shooting a stream of green slime and scoring a direct hit at That Baby's head. Miranda screamed, tossing her stricken doll in the air. Kowhai took one look at the vicious alpaca galloping toward her and beat it, zipping out of the paddock and back to the safety of the house.

We retreated, Miranda wailing and clinging to my leg.

"Did that fucking camel just go for my *brains*?" Peter wanted to know.

"They're not camels," I corrected him. "They're *camelids*."

"My *baby*! I forgot my *baby*!" Miranda wailed.

"I'm not going back in there," I told her. "We'll get you a new baby." Turning to shut the paddock gate, I cast a wary glance back at the boys. *Did I just see what I thought I saw?* They were standing beneath a tree looking cute, chewing their grass and pretending that nothing had happened. "I gotta call Gay," I muttered.

That evening, I rang our breeders. "Oh, don't be silly," Gay chirped lightly over the phone. "Alpacas don't attack people."

"These ones did," I assured her. "They're getting a little aggressive."

"Try to socialize more. They need to get used to you," Gay suggested. "Did you know there's an alpaca in China who predicts sports results? They're very psychic."

So we kept trying. After a couple of weeks Miranda stopped insisting the "mean 'pacas" were trying to "get" her, and we did spend more time with the boys. Usually we did this in the evening, after Peter was home from work and we could go as a group, so we'd be less vulnerable. We'd pour ourselves glasses of wine, fill the purple plastic bowls with alpaca nuts, and wander out to the paddock, hoping for the best.

It did seem that the alpacas were calming down. They'd worked out their ranks according to an uncomfortable racial hierarchy, with Kenny, the large white male, presiding at the top. Henri, the light-brown alpaca, was second in command, and McTavish, the gorgeous chocolate-colored one, got spat on by everybody.

"It's so embarrassing," Peter complained one evening, offering the food bowl to McTavish. Kenny shoved him out of the way and grabbed the fresh nuts for himself. "We thought they were so cute, and now it turns out they're these racist camel zombies."

"Who spit," I added.

"And attack children."

"Oh, well, at least they cure autism!" I reached out to pet Henri's snout. "And . . ."

Peter looked up. "What?"

I blushed. "No. It's embarrassing."

"*What?*"

"Feel Kenny's nose," I suggested. "Does it . . . remind you of anything?"

Peter frowned. "What?"

"A penis," I mumbled, half under my breath.

"A *what?*"

"A penis," I repeated.

Peter snorted, still stroking the nose. "Oh, my God." He laughed. "You're right. I feel kind of gay right now."

And that, I am pleased to reveal, is the sole redeeming quality of the racist camel zombie. Their noses feel exactly like a lovely erect penis. Not a menacing penis, but a friendly penis. The sort of penis you'd like to snuggle up with on a cold winter's day. The penis of your best lover, firm yet warm and accommodating, and covered in a delightful soft fuzz.

"I wish *I* had a furry penis," Peter mused. "I think it would make me more lovable."

"I don't," I told him. "You'd have to shampoo it all the time. Just one more thing to remember."

"G'day," came a voice from behind us. We turned around to see Hamish, the dairy farmer from across the way. Dressed in his usual getup of olive-green coveralls and gumboots, he was leaning on the gate to the alpaca paddock and frowning. "Feeding the llamas, are ya?" he asked.

"Alpacas," Peter corrected, finishing the last of his wine.

"Yeah, we're just feeding them," I chimed in. The fact that we had been performing hand jobs on their noses did not seem necessary to share.

"Was wondering if you had a teat I could get at," Hamish asked, shifting uncomfortably on his feet. "My ewe's had twins you see, and we've only got the one."

At first this seemed a little personal, but then I remembered a ewe is a sheep, and I relaxed.

"I don't know, Hamish. We don't have any lambs." Peter took my empty glass and headed back to the house.

"But we've got some old baby bottles," I ventured. "Let's see if I can find one for you."

Once I set Hamish up with a bottle and teat, I went back to the house. "He's always so stern," I complained to Peter, pouring myself another glass of wine. "Do you think he just can't talk to women?"

Peter grinned. "I don't know, Antonia. Take a look in the mirror."

I did. My T-shirt was smeared with green alpaca slime, I had a glass of white wine in one hand, and I was wearing a pair of red plush devil's horns. If I'd seen myself on the sidewalk of a major city, I probably would have crossed to the other side of the street.

"No wonder the farmers won't talk to us." I shook my head. "We really don't fit in around here, do we?"

That was putting it lightly. Peter and I tried to relate to our farming neighbors as if we spoke the local vernacular, but there was no hiding who we really were: urban Americans who talked funny and knew nothing about life on a farm. Also, we did stupid things like keep alpacas for fun.

Hamish's skepticism about our credentials was perfectly fair. Farming was his livelihood, and we thought it was some kind of a lark. You don't see dairy farmers moving to the city with big ideas about being cardiologists for fun. "How hard could it be?" they might chortle, spitting tobacco and hitching up their jeans. "I'll just git me a book from the library."

Which is more or less what we were doing. Playing at being in the country, like those eighteenth-century oil paintings of Marie Antoinette herding sheep. Not that we were French nobility. We were just another couple of hyper-entitled Americans with liberal arts degrees and a farm dream.

Clearly, I had no credibility with the locals. And for the first time since middle school, I actually wanted to fit in. I liked these friendly people with their backyard sheep and their kids, their easy discussion of chicken bowels and the slime coat on an eel. I didn't understand Hamish, but I respected him. I wanted to learn some of his skills.

"Maybe we should get a real animal," Peter suggested. "We might redeem ourselves that way."

"What about a goat?" I replied. "Amanda said we could just *have* Pearl. We wouldn't need to pay for her or anything." I left out the part about goats being a lot of work, because that seemed like a minor detail.

"A goat, huh?" Peter looked skeptical. "What do you get from a goat?"

"Cheese, of course! Delicious cheese!"

"*Cheese!*" Silas yelled, jumping over to his father. He was wearing a matching red shirt and sweatpants. Hopping up and down with excitement, he looked like a demented elf. "Cheese, peese!" he hollered.

Then he reached out a hand and tried to touch Peter's computer. Peter *hates* this. Children who touch his computer pop letters off the keyboard and smear greasy handprints on the monitor. "Stop it, Silas!" Peter snapped. "That's Papa's computer!"

Silas found this hilarious, and resumed hopping. He pulled a tiny car out of his pocket and then tried to touch Peter's computer with the car.

"*Ha!*" Peter barked. "Not even with the car, Silas!" He put up his hand. "*Stop.*"

Miranda made her entrance then, sporting a purple feather wig and a rainbow bathing suit. As usual, she wore her shiny black gumboots. "Papa!" She strode to her father.

"Yes?" he asked, turning away from Silas, who immediately started pounding the computer.

"Do you want to smell my finger?"

"No."

"It smells like poop!" She placed her finger to her nose, inhaling as though sniffing a fine wine.

Peter looked at me wearily. "What were we talking about?"

"Beats the hell out of me."

"Goats," he said, sighing. "We're getting a goat. I guess we'll call Amanda in the morning."

Eventually we calmed the children down. I washed Miranda's hands and pulled off her gumboots, convincing her that cotton pajamas were more suitable sleepwear than a feathered wig. We tucked in the kids and read them a story, then checked on the alpacas and put the chickens on their roosts. Finally, we lay down to sleep. Our bed was situated beneath a large picture window, and on clear nights we could see thousands of stars, just above our heads.

Peter kissed my neck and reached for my hand in the dark. "Don't worry," he soothed. "We'll find our place here."

"Are you sure? What about Silas?"

"He'll find his place, too. It's good for him out here. Chasing chickens, petting his dog."

I sat up. "Wait a minute. What's that?"

"What's what?"

"There's an alpaca nose on my leg."

Peter pulled me to him. "No, baby. It's not an alpaca. It's just me."

CHAPTER FIVE

SHEEP ON A SPIT

March and April see the end of summer in New Zealand, bringing cloudless days and the beginnings of a crisp autumn wind. Our house had a number of mature trees scattered on the property, and it occurred to me one day that they were laden with fruit. Peaches, quinces, apples, and figs—we picked them by the bushel and could barely keep up with the cascading harvest. These were not the perfect specimens I was accustomed to purchasing from supermarket shelves. Pick an apple, and you were likely to find it hollowed away by ravenous wasps. Quinces fell from the trees, where they rotted, growing swollen and black in the late summer heat. One day I bit into a peach, savoring the sun-ripened flesh, until the pit cracked open and an earwig slithered out.

I dropped the fruit, spraying chewed peach on the deck.

"What's wrong with you?" Peter asked, looking up from the six-foot pole he was sanding. I think he was planning to herd alpacas with it.

"These peaches are full of bugs. And the apples have wasps. And I don't have a clue what to do with the quinces."

"What's a quince?"

"Exactly. I think it's some kind of medieval thing. They made jam with it or something. And wine."

Peter stopped sanding. "You make wine? From a quince?"

"Well, *I* can't. But I'm sure it's not that hard to do."

"Antonia." Peter stared at me intently. "What's our weekly wine budget?"

"I don't know . . . sixty bucks?"

"So start making wine! Think of all the money we'll save! Thousands of dollars each year!" Peter paused to take a breath, his eyes growing manic. "Maybe it could be a business. Quince wines of New Zealand. It's a fantastic business idea."

"Easy now." I held up my hand. "Maybe I should make a bottle first, to see if it's drinkable."

I ran the idea past Autumn. "Sure, lots of people do it," she told me over the phone when I asked. "Cider mostly, but you can make wine with any fruit."

"Don't you need yeast and sugar?"

"Can do. Or you could go basement. Skin makes it all the time. Says he just puts a jug of apple juice on his kitchen bench, lets it sit till it starts to ferment. Then he drinks it till he gets the shits."

"Sorry, who?"

"You haven't met Skin? Aw, he's great. Knows everything there is to know about country life."

"And his name is . . . Skin?"

"Yep. Married to Lish, the lady who drives the school bus. Anyway, he's the one to ask about brewing country wines."

I guess I should have hung up and called Skin, but I have to admit, I was intimidated. First, by the prospect of talking to an actual person named Skin, and then by the notion of a wine that you "drink until you get the shits."

So I got out a book from the library. A few pages in, I learned that wine making is really not hard. As long as you keep your equipment clean, it is actually insanely easy to make some very palatable country wines. Just about anything will ferment into alcohol, including pea pods, ginger, and those sneakers you don't wear anymore.

Well, maybe not the sneakers. But cider making couldn't be easier. You grind up apples with a juicer, then add some sugar and wine yeast. Keep the juice in a covered bucket for a week or two, stirring it every day and leaving it in the sunshine to stay warm. Then you pour it into something called a demijohn, which is a giant glass jug that holds about five gallons of booze. Last, you fit an airlock and wait for a month.

The resulting elixir is dry, delicious, and completely deadly. There's a mathematical formula you can use to calculate the alcohol content of homemade wines, but after one glass I was too drunk to add and stopped trying. Within a month, I was brewing gallons of the stuff—on the back deck, in the bathtub, and in the pantry. Buckets of cider and wine were spilling out of the closets and crowding the corridors. The children tripped over them when they walked into the kitchen, searching in vain for a healthy snack. I hardly had time to go to the grocery store, and when I did, it was only to bring home twelve pounds of sugar and loose tea for tannins.

One day, I snapped at the kids. "Where are the raisins?" I demanded.

"*Away*," Silas announced. He was driving a toy car along the kitchen counter.

"You used them all to make booze," Miranda said with the wide-eyed clarity that only a hungry three-year-old can muster.

We brewed hard apple cider; quince and fig wines; beet, ginger, banana, and apple wines—but my favorite was the peach. This was partly because it tasted like peaches, but mostly because it lifted me up on feathery angel fingers and flew me to the Land of Enchantment. Then it dumped me there until Monday morning, when my kids looked dirty and the weekend was gone.

The difficulty with homemade wines is that they're strong, they're free, and they don't give you a hangover. I always thought the headache was a consequence of drinking too much alcohol, because it's morally wrong to have too much fun. But it's the chemicals in commercial wine that make us sick, not the booze. My wines contained fresh fruit, water, sugar, and yeast, and they never gave anyone a hangover. Essentially, they were intoxicating happy drinks that grew on trees and had no consequences.

Except our being drunk all the time, which is a consequence itself if you're trying to live a life that's based in reality.

"Where are my pants?" I demanded the morning after an especially exuberant night of quince wine.

"Don't you remember?" Peter responded with a smile. "You took them off so you could scramble up the water tank to pick figs in your underwear."

"Ah, yes." I smiled, reaching for the coffee. "That was sensible."

Peter's face clouded with concern. "Antonia? Maybe we should share some of this wine with our friends."

"Why?" I demanded, sucking back my coffee. "I need it. I'm thirsty."

"Well . . ." Peter considered. "After the third bottle of cider, we tend to lose track of the kids."

"So? They learn independence. They get to survive in the wild."

Peter ignored this. "Why don't we have a party? Meet some of the locals out here. Give 'em a taste of our hooch."

"Fine," I said, scowling. "As long as it's just a taste. I don't want them cleaning me out of peach wine."

I was going to invite people for a barbecue, but Autumn had a better idea. "Get Skin to do sheep on a spit," she told me Tuesday when she dropped by for coffee and a chat. "Best sheep you'll ever eat. He'll roast that beast all day."

Still feeling anxious about a person named Skin, especially one who roasts dead things on spits, I introduced myself to Lish, the school bus driver. With flowing Polynesian hair and a wide, warm smile, she was easy to approach. I clipped Silas in his seat belt and hesitated, then finally asked, "Lish, right? And your partner's name is . . . Skin?"

"Yep." She nodded. "Too bad for me, the mongrel."

The sparkle in her eyes told me she was joking. "Do you think he would mind cooking a sheep for us? Like if we have a party on Saturday?"

"Sure." She grinned. "No worries. Glad to help."

"Should I . . . how much does he charge?"

"Ah." She waved me away. "Just flick him a case of bourbon and Coke. He'll be happy as!"

"Great," I said, somewhat bewildered. "Saturday, then. If it's okay, just tell him to come round whenever." I slid the bus door shut, blowing a kiss to my son.

"I'm not completely sure," I reported to Peter that night, "but I think a person named Skin is bringing a dead sheep to our house on Saturday."

Peter raised one eyebrow and waited.

"He's cooking it. For the party. For bourbon."

"What's bourbon?" Miranda wanted to know.

"It's a drink," Peter told her. "A grown-up drink." To me, he said, "Do I have to dig a hole? Build a fire?"

"No. Apparently he kills the sheep today, then he guts it and lets it hang for a couple of days, then he shows up on Saturday with the gear."

"May I please have some bourbon?" Miranda inquired.

Silas was engrossed in reconstructing the pieces of his watermelon rind, which he'd formed into a perfect half circle. "Ba," he commented.

"That's right!" I stroked his head affectionately. "The sheep says baa."

"Until we shove a pole up its ass and roast it on hot coals," Peter clarified.

"Seriously?" I shot him a look. "Was that necessary?"

"No," he conceded. "I guess not."

What was necessary was locking up the alpacas, because children were coming to this gathering, and we'd already let our German shepherd chew on one of them. So early on Saturday morning, Peter and I ventured into the paddock where Kenny, Henri, and McTavish were peacefully chewing grass and pretending to be cute and not evil. The plan was for Peter to back them into a corner, where I could easily slip on their harnesses.

"Just follow me," Peter urged. "I've got my lance."

"You mean the stick?"

"It's a lance. I'll take one of these camels out if I have to."

Doubtful, I looked at the stick. "Let's hope it doesn't come to that."

We edged forward, Peter using his stick to separate Kenny from the others. "*Beeeeehn . . .*" Kenny warned, working a wad of green slime in his mouth. "*Beeeeehn . . .*"

"Oh, shut up," I told him, slipping the harness over his snout. "Aren't you supposed to hum? And cure the lame?"

We'd successfully harnessed the alpacas and were tethering them to three wooden posts in the back paddock when a gold sedan rolled down our driveway. There appeared to be an old, rusted oil drum strapped to its roof.

"That'll be Skin," Peter muttered, and I noticed he didn't put down the stick.

A slight, wiry guy unfolded himself from the driver's side and started untying the oil drum from the roof of his car. Getting closer, I saw he had a mess of dreadlocks coming out the back of a well-worn beanie. His skin was brown and creased, with a patchy gray beard and moustache across the lower half of his face.

"Howzit?" He grinned. "I'm Skin." He tilted his head, and I caught a glimpse of his eyes, which were strangely out of place on his scraggly face. They were dark brown, soft, and gentle.

"Let me give you a hand with that drum," Peter offered, and then walked around to the far side of the car to help Skin lift it. He'd put down his stick, but I was tempted to pick it back up again. Despite the nice eyes, this guy didn't look safe. I made a mental note of where my kids were.

Skin lifted his arms for the oil drum, and when he did, I saw the buck knife, snug in an old leather sheath, strapped to the side of his belt.

"Where you want her?" he asked me, then nodded to Peter and winked. "Best to ask the missus these things, I reckon."

"Er, over there by the garage, I guess," I stammered. "Does it really take all day to cook?"

"Takes a wee while," he said, nodding, and snapped open the lid of his trunk. "She's a big 'un."

With that, he lifted out a bald sheep carcass the size of a six-year-old child and flipped it over his shoulder as though it were nothing but a big bunch of flowers. He pulled a five-foot steel pole out of the car, then laid the sheep on the grass and lifted two of its legs up. With a moist crunch, he slid that pole through the carcass until it came out the other side, wet and glistening.

"I think I'll go back in the house," I volunteered. "Check on the kids."

I stayed clear of Skin for most of the day, making salads and chasing my kids inside. He sat out on a deck chair with Phoenix, petting the dog's shaggy head, sipping the drinks that Peter brought him, and turning the crank handle on the side of his drum, which I now knew had been made for molasses. By evening, the sheep's flesh was a taut, shiny mahogany, and the steam wafting toward the house made my mouth water.

Our friends seemed to arrive all at once: Amanda and Nick in their silver minivan, Autumn and Patrice in their old Toyota pickup. Both these families had three kids apiece, and soon the children were swarming, tearing around the property with bare feet, helping themselves to fruit juice, and pedaling the kids' bikes at top speed. Titou and Miranda swiped a party-size bag of Doritos from the kitchen, then clambered up into the totara tree with their plunder.

Maria stepped out of a dark blue pickup truck and sniffed the air appreciatively. She was wearing navy capris that showed off her splendid legs. "Wasn't sure what to think when you asked us for sheep on a spit," she commented. "Means something else in England."

"What's that?" Amanda asked.

"Two blokes on one lady," she explained with a ribald wink. "Getting done at both ends." She poked two fingers in the air to illustrate.

Children started screeching behind us. I turned in alarm, scanning the crowd for broken bones.

"*Skin!* It's *Skin!*" Sophie and Amelia hollered, pouring out of the van and tearing past us toward the sheep spit, where the fathers were standing and chatting. The two girls crashed headlong into Skin's legs, and he scooped them up, flinging them over his shoulders like wiggling sheep carcasses. Lucy staggered behind her big sisters, holding out her arms to be lifted. Even Silas was delighted, flapping his hands and grinning at the uproar.

"Oh, God. I thought they were screaming," I said to Amanda. "Are they scared of him?"

Amanda shook her head and laughed. "Not Skin," she said. "Kids adore him. He's like the Pied Piper of Purua."

Sophie and Amelia were now trying to scale this man like a tree. Though his frame was small, Skin had arms as strong as steel cables. He lifted each girl with one hand and allowed her to walk up the front of his legs with no strain at all. Then he dangled her out over thin air, squealing and giggling, before depositing her gently on the driveway.

"He's like the Grim Reaper." Autumn had come up behind us. "Anything you need killed, he'll do it. That man's like a cockroach. He'll be here long after the rest of us are gone."

"Autumn!" Abi chided. "That's a rude thing to say."

Autumn reached for a glass of cider. "It's a compliment, really. He can kill and butcher. He can fish; he hunts pigs. When the zombie apocalypse happens, I'm sticking with Skin."

Maris ran by, chasing Nova, and I noticed she had a blood-soaked gash dripping down the side of one arm. Her neck also appeared to be slit. "Nice work, Maris!" I called, pointing at the gore. She grinned back and waved, taking off in the direction of the trampoline.

"Why are the alpacas tied up?" Amanda asked, peering toward the back of the paddock. "I thought maybe the children could go out and visit them."

"*No!*" Peter yelled, then coughed when everyone looked concerned. "I mean, we thought it might not be safe."

"It's just . . ." I stammered. "I think they want to eat people's brains."

Amanda looked as though she had something to say about this revelation, but she stopped short, as Sophia had just arrived with her boyfriend, Bill, in a silver Mercedes. She stepped out of the passenger seat dressed entirely in snowy white linen. Bill, an older businessman from Auckland, parked the car and took her arm. The silver cufflinks on his crisp white dress shirt glinted in the sun.

"Thank you, you sweet thing," Sophia said as she took a glass of peach wine from Peter. "I'm simply gasping."

"You're not serious about the alpacas," Amanda said, nudging me. "Look at them! They're such gentle animals!"

I shrugged. "We can go out there, if you want. But I wouldn't take the kids if I were you. Alpacas are highly unpredictable."

"I had the most ghastly week," Sophia volunteered, wilting onto a deck chair.

Abi plopped down beside her. "Oh, no! What happened?"

"Duncan," Sophia said ruefully, then took a sip of her wine.

Amanda's voice turned serious. "But Michiko's on holiday. Was he up at the school?"

"He made an appearance. I had to do a lockdown procedure with the children. I shut them in the classroom and made them hide under their desks whilst I got him off the property."

"Wait," I interjected, alarmed. "What? At Purua School?"

"It's Michiko's husband," Autumn explained. "His name's Duncan. He's—what is he, Sophia? Manic-depressive?"

"The guy's nuts," Bill interrupted. "He's nuts, and I keep telling her she should call the cops."

"He's . . . rather unstable," Sophia smoothed. "And he's fine when he's on his medication, but when he goes off, he's very unpredictable, and he makes these awful threats against Michiko. Then he comes to the school and rants at the children."

This gave me pause. I hadn't anticipated mental illness or lockdown procedures in the Shire. "He's not dangerous, is he?"

Sophia shook her head. "He'd never strike the children. He just can't modulate his behavior. He uses inappropriate language—"

"Told one student to fuck off, as I recall," Amanda interjected.

"—or he yells, just acts erratic. And the children are entirely my responsibility when I'm out there. I can't be too careful."

"So call the cops," Bill repeated irritably. "That's what they're there for."

"But Michiko seems like such a gentle lady," I protested. "I can't believe she's got a violent husband."

"Every community has its secrets," Sophia said dryly. "Come to Purua. Runny scrummy honey. Best drunken parties. Only slightly insane."

"We're all a bit nutty to be out here," Abi agreed. "You'd have to be, really. It's not like any of us knows what we're doing, with the animals and this country life. I can't even cope with my chickens."

Bill tipped his glass at Sophia. "That's what I tell the woman.

Why live in the sticks when you can come to the bright lights of Auckland?"

Autumn still looked worried. "So, was it all right?" she asked. "You got Duncan off the school grounds?"

"Of course." Sophia waved her glass dismissively. "He wasn't making any sense. He just needed a firm hand, that's all."

"Sheep's done," Peter announced, coming over from the spit. "Skin's carving it up now."

A single sheep produces a formidable pile of meat, enough to feed the whole community. Skin carved up the roast with his buck knife, and the way he glided it through the flesh, it was clear that blade was as sharp as a surgeon's scalpel. There were slices of haunch, big chunks of backbone dripping with fat, and drumsticks so huge they looked Jurassic. The children came running when they saw the meat, the older ones helping themselves to plates, the younger ones hanging on their parents, demanding to be served. Kowhai and Phoenix stood around hopefully, wagging their tails and looking mournful. Skin tossed them a handful of scraps, and they fell on the meat in rapture.

"Mama, look what Maris did!" Miranda yelled, holding up her doll for inspection. "Isn't that cool?" That Baby appeared to have been the victim of a slasher attack. Her plastic throat was slit and her belly eviscerated. The doll was soaked in ketchup, a spoonful of canned spaghetti standing in for her entrails.

"That's very nice, Miranda." I smiled, feeling ill. "Would you like something to eat?"

The sheep, it turned out, was delectable. The meat was meltingly tender, each bite suffused with fat like a fine marbled steak. "Yep, that lamb you get in the shop's no good," Skin explained. "Kill it too early, before it's had a chance to develop the flavor. This one's done proper, all the grease and the fat still in."

"Done at both ends." Maria raised her glass. "To perfection!"

"Thank you, Skin," I told him, this time with a smile. "The meat is incredible. It's some of the best I've ever had. Do—I mean, do you think you might teach me how to cook a sheep sometime? I'd like to try it."

He looked pleased. "Yep, we could do that."

But Lish just shook her head. "He'll never tell. Moans all day about the work it takes, but he loves it. Wouldn't give it up for the world."

Skin laughed and shoved her arm with affection.

After our meal, we sat under the stars sipping the ninth or tenth bottle of peach wine. Or maybe it was cider. Or maybe it was both. By that time, I didn't much care.

CHAPTER SIX

STRIPPER CALVES WITH SATAN TONGUES

Following their lurid performance at the party, I decided that Nova and Maris needed a better palette to draw from. I poked around on the Internet and found a theatrical makeup supplier in Hamilton.

Later that week, I ran into Autumn at the school. "I have sort of a strange question for you," I began.

Autumn waited.

"Would you mind if I gave your children blood?"

She did a double take. "Pardon?"

Thankfully, it wasn't too awkward to explain, and in the next few days, I put together a little care package for the girls: a dozen blood capsules, a package of wound wax for creating realistic effects, and half a liter of stage blood, in a purplish, arterial hue. I dropped the package on their doorstep, feeling good about playing Secret Santa, and promptly forgot about it.

And that's when things got really dark. One night in March,

Abi woke to a strange sound in her house. "It was like a bowl of Rice Bubbles," she told me. "You know, the cereal? With Snap, Crackle, and Pop? Just a real soft popping noise."

She padded out into the lounge to see where the sound was coming from. It should have been pitch-black at two in the morning, but the room was lit up with a pale yellow glow. When she drew back the curtains, she saw Michiko's house, just across the road—or rather, what was left of it. The house was consumed in flames.

No one was home when it happened. Michiko and her kids were on their way back from a holiday abroad. When Michiko got the news the next day, she moved away to stay with friends. Her home, clothing, furniture—everything she owned was in ashes. All she had was her kids and their bags, still packed from their fun-filled holiday.

For the next few days, everyone talked about the fire. We all wondered how it had happened. Was it an electrical fault? Or maybe a broken hot water cylinder? But it didn't take long for people to start whispering about the husband.

I told Peter the news about the fire. "And that's not the worst thing," I told him. "People think it might be arson."

Miranda looked confused. *"Our* son set the house on fire?"

"Do they know who did it?" Peter asked.

"No, there's no evidence yet. But her husband was always making threats."

"Did Silas burn the house?" Miranda was still trying to puzzle out the mystery.

"No, Magnolia," I corrected. *"Arson.* It means somebody burned the house down on purpose."

"Oh, okay." Miranda thought about it. "Arson. That's good."

"Is Michiko okay?" Peter asked. "Can we bring her anything?"

"No." I shook my head. "She's in hiding, staying with friends or something. She's going to change her name. We're probably not gonna see her again."

That put a pall on the day. I started washing the breakfast dishes while Peter took the kids out to ride bikes, but I couldn't stop thinking about the fire. The green hills outside our kitchen window were lit up with sunshine, and I could hear my kids shrieking happily outside with their father, but it all seemed so fragile now. Was this where we wanted to live? A place where houses burned under suspicious circumstances? Where people were scared of their neighbors? And where was this husband now? Would he come after the rest of us?

When I lived in the San Francisco Bay Area, I berthed my sailboat in Richmond because it was the only place where I could afford a slip. Even in the boom years, Richmond was poor and shabby, plagued by drugs and gang violence. Peter and I used to lie awake at night listening to gunshots in the streets. I taught for a while at the local high school, and I remember being shocked to see the heavy gray metal detectors guarding the front entrance, the walls scarred with decades of graffiti. Once we moved to New Zealand, I thought we'd left all that behind. There weren't any wars here. New Zealand wouldn't even let nuclear subs in its territorial waters. Purua's charming one-room schoolhouse, the cute kids who made their own honey—where did arson and mental illness fit into that mix?

I wasn't worried just about the four of us. Peter's niece Rebecca was coming to live with us soon, a shy, introverted twenty-year-old who was studying fiber arts in college. Though she'd been raised in suburban New York, Rebecca was a placid vegetarian who liked petting cows and spinning wool. A perfect evening for Becca might consist of a mug of hot chamomile tea and a nice sock-knitting project.

I dried my hands on a tea towel and called Amanda. She answered the phone on the first ring.

"Look, Amanda. I heard about the fire, and . . . I'd like to help." It seemed weird to say, since we'd been in this community only a couple of months. "I know I just met Michiko that one time, but—"

She answered before I'd even finished. "Absolutely. Karl rang me earlier today. He's doing a collection over at the school—anything you can spare. Clothes, food, furniture, gift cards. She lost all her family photographs, too, so I'm making an album with pictures of her and her kids."

"But I thought she went into hiding. How does he know where she is?"

"He doesn't. No one does. But we can leave stuff with her friends in town. We have to help her; she's got nothing. Oh, and by the way, did you still want that goat? Because you can come and get Pearl whenever."

"Yeah, yes. Definitely."

She paused. I had the impression Amanda was jotting something down. "What do you want to do with her, anyway?"

"Knock her up. I mean, not personally, but, you know. Find a ram."

"A buck."

"You want money? I can pay you."

"No." Amanda laughed. "You'll need to get a *buck* on her. A ram is a sheep. You won't get far with that one."

Clearly, I had a lot to learn. I said my good-byes and hung up the phone before I could disgrace myself further.

The next day was Saturday, and I drove to the school with a gift certificate from our local big-box store, thinking that Michiko might want the basics: new toothbrushes, clean shirts, fresh pajamas. On the way back, I planned to stop at Amanda's. My backseat was folded down, the rear of the station wagon lined with old feed bags. I was picking up our new goat.

And that's how Pearl trotted into our lives. Her luggage consisted of a chain and two steel posts that screwed into the ground. "You have to use two," Amanda explained. "If you stake her with one, she'll pull it right out."

"I just thought I'd let her roam free," I suggested, with the optimism of someone who's never had a goat. "She can browse her way through the paddocks."

"That's really stupid." Amanda's eyebrows sparred like two angry caterpillars. "She'll eat your garden and then she'll eat your trees. She'll eat everything she finds, if you let her. She might even take a bite off your kids."

A flock of chickens shuffled past, pecking and scratching at the grass. A tall white rooster with a black ruff of feathers at his throat strode among them. He threw his head back and let loose a vigorous crow.

"You have a rooster," I observed pointlessly.

"Yep, that's Goldie." Amanda crossed her arms in front of her chest and nodded in his direction. "We'll have to find a home for him soon."

"You don't want him? I'll take him." Idyllic farm visions flashed through my head. Dozens of tiny chicks, adorable little puffballs that tumbled on my children and tickled their tummies. The sound of a rooster's crow in the morning, just like we heard in our years of sailing, when we'd anchor near small villages in Mexico and Central America.

Amanda said something, but I'd missed it. "What?"

"I said, 'You can have him.' No problem. You can take him today, if you want."

"No, that's all right. I have a lot on my plate with the goat. Besides, I don't have anything to carry him in."

But Amanda was already hustling toward the garage. "No worries. I've got an old cat carrier you can use. Hang on one sec while I dig it out."

I should probably have asked why Amanda was so eager to give away her rooster, why she stuffed it in a crate and tossed it in the back of my car, then waved good-bye with such a cheerful smile on her face. But I didn't. Instead, I just drove home, too excited about our new goat to think about much else.

We staked Pearl with the two steel posts right outside our window, and she seemed relaxed in the comforting shade of the palm tree. In the mornings we'd drink our coffee, watching her as she got to know our other animals. Kowhai liked to play with Pearl, and by play, I mean "chew on her leg." When she spied the dog, Pearl's ears went on high alert. If Kowhai got closer, she'd rear up, her ears standing straight up like devil's horns.

Toward the middle of May, Rebecca arrived in New Zealand. Her plan was to spend eight months with us, taking the fall term off, and to return to college in January. Peter drove to town one Saturday to pick her up, and when he came back with her, she was just as I remembered: a hippie pastiche of ankle bracelets, bare feet, and organic hemp chic.

Tall and slender, with long dark hair and a smattering of freckles across her face, Rebecca is seriously into the natural look. No makeup, no hair products. Embroidered jeans. Moccasins. Clothing purchased by the pound from a thrift store. And she'd flown to New Zealand with her own wooden spinning wheel.

"Is this, uh, everything?" I asked, helping to lift her medieval contraption from the trunk.

Rebecca giggled. "Oh no, I have a suitcase," she said, pulling out a small bag on wheels. "I don't need very much stuff."

"Becca!" Miranda called, shooting out of the house. "You did fly from *America*?" Though she'd met Rebecca before, when she was a baby, Miranda couldn't have remembered her. Nonetheless, she was ecstatic to have a brand-new playmate.

Rebecca knelt down to talk to her. "That's right," she told her solemnly. "And I took *five airplanes* to get here."

Miranda's eyes grew wide. "Five airplanes?" she repeated. "That's amazing!" I'm not sure Miranda even knew what an airplane was. Nonetheless, at three, she was ready to be impressed.

Silas appeared on the deck, his Dart pressed to his ear. He smiled shyly at Rebecca. When she caught sight of him, she scooped him up and hugged him, and over her shoulder I could see a wide, happy grin on his face. *He probably does remember her*, I reflected. He was three years old the last time she visited.

For the rest of the afternoon, the children were glued to Rebecca. She went out to meet the alpacas and visit the goat, and they followed her to the paddocks. She sat down to eat the spanakopita and salad I'd prepared for lunch, and the children warred over who would sit on her lap. She flopped on the couch to tell us her family's news, and the children sat on either side of her, pressed against her skin. She looked like some mythological hippie monster staggering around with embroidered jeans, six arms, and three heads.

And she didn't have just a spinning wheel with her. She'd also brought her iPod. "Have you heard the 'Harlem Shake'?" she asked Miranda, plugging her gadget into our speakers. A pulsing dance beat filled the room, and Silas started hopping up and down while

Miranda merely stared in open-mouthed amazement. Then Rebecca plugged into our wireless Internet connection and pulled up all the viral videos we'd missed while raising kids in New Zealand: a killer snowman, the ghost of a dead girl in an elevator, and all the newest hip-hop videos. By the end of the night, Miranda was wearing a hot pink bathing suit and rotating her rear end like a gifted stripper.

"Booty wurk, booty wurk," she sang along with the video. "Left cheek, right cheek! Left cheek, right cheek!"

I cast a worried glance at Peter.

"It's cultural!" he protested. "T-Pain's teaching her left from right!"

Toward eight o'clock, Rebecca yawned and stretched, pulling her faded hoodie on over her T-shirt. "I gotta go," she sighed. "I'm wrecked."

Miranda was alarmed. "Are you going back to America?" she demanded. "Will you take five airplanes?"

"No, honey." Rebecca ruffled her hair. "I'm just going to the sleep-out. That's where I'll sleep while I'm here."

"*Away*," Silas volunteered.

"That's right." Rebecca gave him a squeeze. "I'll be back in the morning."

The sleep-out was a shed we'd fixed up behind the house so Rebecca could have some privacy while she stayed with us. But our efforts were hopelessly optimistic. The next day was Sunday, and Miranda woke up before dawn and padded through the wet grass in bare feet to wake up Rebecca in her not-so-private room.

Peter and I lay in bed, grateful for the chance to sleep in. From our window, we could see Pearl rising up on her hind legs, nibbling the palm fronds from the bottom of the tree. "I think I love our goat," Peter observed. "I mean, I really have a *relationship* with her."

And it was true. Each day, when he got home from work, he'd bend down and press his head into Pearl's. She'd press back with surprising strength. Peter didn't always win.

But our affection for Pearl was tempered with a note of anxiety. After the chicken leprosy fiasco and the unfortunate incident with the leg mites, I was determined to do some research on our first real farm animal. I checked out every goat book from the local library, and what I learned was somewhat disturbing.

Take mating, for example. The male goat, or "buck," as he is called, is a known sex maniac. He can be "in rut," as the period of sexual arousal is known, *for months on end*. During this time, he will groom himself for romance by repeatedly urinating in his beard until he is surrounded by a miasma of pee-soaked goaty funk. Then, just to be thorough, he pees in his own mouth and blows himself.

On top of this, he is horny. All the time. And by horny, I don't mean that he'll gently rub up on you to elicit affection. I mean that if you're a goat keeper, you're advised to bring *Tasers* and *pepper spray* into the buck pen, just in case you turn your back and that buck tries to ass-rape you from behind.

Now, Peter and I are gentle people. We're both left-wing, gourmet foodies. We like local food, slow food, and most things involving arugula. Since moving to the country, we occasionally read the urban farming blogs, even the wide-eyed ones with no ducks or maggots. "Our goats are so enchanting," a typical entry will brag. "Once you've seen your baby buckling prancing about with the family dog, you'll love these little imps for life." And sure, that sounds nice. But no one likes to talk about how that cute baby buckling will grow up to be a caprine sex fiend who rushes you in the barn while spraying clouds of noxious goat piss in the air.

I thought they just made cheese.

So although we were eager to produce some enchanting little baby goats, the whole process of getting Pearl pregnant put me off a little. Serious goat breeders hold the female down while the male checks her out. Sometimes she'll pee on him, and the buck will lap it up, because that's super sexy. Then, after the mating is complete, the buck will stamp around looking satisfied, and everyone will go home feeling troubled and smelling of Satan.

The alternative is to purchase a straw of frozen goat semen, which is thawed and shot up your goat's vagina by a professional, but as soon as I read the phrase *frozen goat semen*, I had to stop reading and think about butterflies.

So we went for option three. We sent Pearl to Love Mountain, a hillside owned by Richard and Jackie, our neighbors down the road. Rumor had it, that land was overrun by a herd of randy, semi-wild goats. Pearl would undoubtedly be gang-raped, with all the ensuing unpleasantness, but at least we wouldn't have to watch.

"How are we going to get her in the car?" I asked, peering doubtfully at Pearl's powerful frame. When we'd picked her up the first time, Amanda had been in charge. But in the few weeks we'd had her, it had become abundantly clear that Pearl went only where she chose to. Just once, I'd hooked a lead to her collar and yanked her in my direction, and it seemed she had the power to tether her body to the center of the earth. No matter how hard I struggled, she refused to budge.

"Oh, that's easy." Rebecca smiled and pulled out a shopping bag. "We'll just give her banana peels. She loves them."

Rebecca dangled a limp, brown peel in front of Pearl. Her ears stood on end, and when Rebecca tossed the treat into the back of the station wagon, the goat hopped in there like a happy golden retriever.

"How . . . how did you know that?" I asked, impressed. "And where did the banana peels come from?"

"Oh." Rebecca shrugged. "I fed her a peel one day, and she just went nuts. So I've been saving them out of the garbage."

Okay, I thought. *That's normal.* Now that Pearl was safely contained in the back of the car, we pulled out onto the road and drove the short way to Love Mountain. We tempted her out of the back with a bouquet of banana peels, then unlocked the gate and coaxed her inside. She followed us docilely until we got halfway up the hill and she caught a glimpse of the wild goat herd.

Pearl stopped in her tracks. She bleated softly. I don't speak goat, but I'm fairly confident she'd said, "Holy hell. You people aren't *leaving* me here, are you?"

And then we tried to walk away. Increasingly anxious, Pearl followed us. Eventually, Rebecca and I had to lock her in an inner paddock before hurrying back to the car, closing our ears to the rising sound of panic in her screams. Before I got in, I cast one last glance up the hill. Those bucks were advancing. They had something to give her, and I'm pretty sure it wasn't banana peels.

Once Pearl was away on Love Mountain, we had more time to spend with our rooster. In the first few weeks, he'd seemed pleasant enough. He herded our young hens around the garden, urging them to eat before he did and crowing lustily, pleased at his own magnificence. He was about eighteen inches tall, with clean white feathers and a vigorous red comb. We all agreed he was a handsome bird.

But as he got older, Goldie started to change. Maybe it was puberty, or else he had his own brand of manic depression, but that rooster turned into a first-class cock. I went down one morning to feed the hens, and the next thing I knew, a squawking bundle of rage hit my leg. "What the *hell*, Goldie!" I screamed. "I'm *trying* to feed you!"

It's a funny thing that we call them cocks, because the rooster doesn't have a penis. Unlike the duck, with his devilish member, the rooster has a male version of the cloaca, which he touches to the hen's cloaca in a gesture called the "cloacal kiss." This sounds sweet, and even romantic, but if you've ever seen a rooster mating, you know better. He launches himself at the hen, pins her down on the ground, flutters, and then leaps away. It's not a courtship so much as an assault. It's over so fast the hen doesn't know what hit her.

And Goldie didn't break for breakfast. Whether it was my leg or the chickens, that rooster was fighting or fucking all day long. Or else crowing, which started out charming but got old fast. In no time at all, it was clear this bird was hopelessly misnamed. One morning, I was reading a Lewis Carroll poem to the children when his true nature came to light.

"I'm changing his name," I announced that night at dinner. "That rooster's name is Jabberwocky."

"Isn't that a monster?" Peter inquired. "With jaws that bite and claws that catch?"

"Yep." I nodded, slicing a bite off my steak.

"And don't they cut its head off in the end?"

"Exactly."

"Jabberwocky seems like a good name to me," Rebecca piped up. "That rooster's an asshole."

May gave way to June, and as we approached the halfway point in our year-long rental, we were forced to confront the inevitable: unless we found another place to live, we'd have to move back to town. And I didn't want to leave. We had friends now and a community where we wanted to live. Beautiful Autumn and her French husband, Patrice; bewildered Abi with her problematic chickens; the elegant Sophia. Even Amanda and her dangerous husband, Nick, were dear to us now. For the first time in years, we felt that we were home.

And yet our rental was for just one year. Katya and Derek were returning in January, and we were pretty sure they'd want their house back.

"Maybe we could hide behind the furniture," I suggested, eyeing the oversize couch. "We could come out at night and nibble on scraps from the kitchen. They might just think they have a rat problem."

Peter shot me a look. "That's not practical."

There was nothing for it. We had to look for a house of our own. But this was a tougher proposition than you might suppose. As I've mentioned, this part of New Zealand was settled by a few large farming families, and their descendants owned massive tracts of land, sometimes hundreds of acres, often worth more than a million dollars. Even if we had that kind of money, these were serious farms, and we wouldn't know how to run them. Raise beef cattle? Be sheep farmers? I didn't know the first thing about either of those projects. I couldn't even keep my horny rooster under control.

"You could run a dairy farm," Rebecca suggested. "I could help you." This was less of a shock than it might at first seem. Rebecca genuinely knew a great deal about farming. Never a fantastic student, halfway through high school she had put her foot down and refused to go back to class. Peter's sister Susanne, Becca's mom, had been beside herself. Like an angry goat with her hooves dug in, Rebecca refused to go where she was told. Her parents tried to coax her back to school, then bribe her, then force her, then physically drag her. Nothing had worked. Finally, they threw up their hands. And sent her to the Putney School.

Located on a working farm in rural Vermont, Putney is an exclusive private boarding school where, in addition to their class work, students are taught responsibility and real-life skills by caring for livestock and running the barns. The school produces a steady stream of high-end milk and butter and sells these products for money. The business plan is genius: they hire child laborers, then charge them fifty thousand dollars a year to work for free.

That's my bitter and cynical take on things, but the truth is that Putney saved Rebecca. Instead of dropping out of high school, she thrived at Putney. By her senior year she was "barn head," which meant she ran a crew of eight to ten student workers, banking a small stipend for the position. In addition to earning her diploma, she learned how to care for animals, manage a staff, and run a small business.

But despite Becca's enthusiasm, I couldn't see us running a dairy farm. Instead, I scoured the real estate listings for "lifestyle blocks," smaller parcels of land with a house on just a few acres.

A few possibilities turned up. Our favorite property was sixty acres of native New Zealand bush, where a Slovakian architect and his wife had built a pretty little three-bedroom home run entirely on solar power. The land was at the end of an unmarked road and it had no address, but so what? The pictures we saw on the Internet were lush and green, like pages torn from *National Geographic*.

A nervous real estate agent named Kim drove us out there, chatting frantically about the charms of the New Zealand countryside. She almost made us forget we were traveling down steep gravel switchbacks, homemade roads that were sure to flood and wash out in the first serious rainstorm.

Then we met the vendor. Martin was a mountainous man with wild hair and a generous belly hanging over tattered black shorts.

The reason there was no address, he informed us with pride, was that he'd built the house and outbuildings illegally. Not a stick of it had any consent from the city council. It didn't even know the place existed.

"Oh," I said tentatively. "I think that might be a problem."

"*Problem?*" Martin roared. "Why '*problem*'?" We knew he'd had some complications from diabetes and that's why he was selling the land. I wondered if all his medical problems had affected his hearing. Everything he said was in a yell. He reminded me of an enraged bear.

"Well, I'm just guessing here," I ventured, "but it might be hard to insure a property with no building consents."

"*Why* do you want insurance?" he rumbled. "You want insurance, go live in the suburbs. This is the *bush*."

It turned out that *Go live in the suburbs* was a favorite phrase of Martin's. When we asked about the lack of heat or a telephone, Martin told us to "Go live in the suburbs." When we expressed a gentle skepticism that a dishwasher could actually be run with power from a windmill, Martin invited us to "Go live in the suburbs." Through all of this, his wife, Klara, a diminutive woman of very few words, sat nearby and cringed. Occasionally she filled her husband's glass with apple juice.

The house itself was pretty, built with richly varnished native woods, with wide picture windows and a wraparound deck. But it was hidden, not just from the law but literally. It was tucked in an overgrown valley, at the bottom of the death-trap driveway, then off to the side, where it would never be exposed. Martin seemed edgy and threatened by something more palpable and reality-based than our jokes about the zombie apocalypse. The property made me nervous.

"I don't think he wants to sell," Peter commented when we

finally escaped to drive up the unmarked gravel road. "He didn't really seem to want us around."

"That's for sure," I agreed. "What's he running from back in Slovakia? Who leaves their home country to go live in New Zealand and hide in the woods?"

Peter rolled his eyes at me, which I chose to ignore. "Anyway," I continued, "how does he think he's going to get half a million dollars for a house with no power and no heat?"

"Well," Peter said, shrugging, "you're in the deep country now. There're some strange people out here. If you want normal, go live in the suburbs."

But I wanted to live right here, in Purua. And Peter must have felt the same way, because the next thing I knew, he was talking about getting more animals.

"Cows," he announced one morning. "We need cows."

"We do?"

He put down his tablet. "Pearl isn't enough, and anyway, she's up on Love Mountain. We need something that . . . makes something."

"I can make something!" Miranda entered the room balancing a small pile of fingernail clippings on a saucer. "Look, Papa! I made you a cake!"

Rebecca wandered in, still dressed in her pajamas. These consisted of a tattered T-shirt, black sweat pants, and thick rainbow leg warmers made from hand-dyed wool. Sleepily, she wandered to the kitchen and started brewing some chamomile tea.

"Thank you, honey." Peter took the fingernails from Miranda and pocketed them before they could scatter on the living room rug. "I was thinking something more like . . . a cow. They make milk."

Silas looked up from the monochromatic towers he was building with Legos, each one sorted by color. "Mih! Mih peese!"

I got up to pour him some. "Has it occurred to you," I pointed out, "that we just got rid of a cow? Lucky was a complete disaster. She practically got our daughter killed. Plus it's her fault my car smells like poo."

"Lucky pooped in your car?" Rebecca asked. "How did she even fit in there?"

"Not Lucky." I shook my head. "Silas. But I still blame the cow. She kept jumping fences."

Rebecca flopped down on the window seat and warmed her hands on her mug. "She was probably lonely."

"*Seriously?*"

"Well, yeah. I'm sure she can see Hamish's cows across the way. Cows don't like being alone."

I sat down on the couch and rubbed my temples. "Okay. So Silas gets speech therapy, Miranda goes to swim lessons, and now what—playdates for a cow?"

But Peter was already nodding his head. "That makes a lot of sense. Cows are herd animals. Like alpacas, like people. They need to be with their friends. I bet if we get two calves, we won't have the same problem."

"So now we need two cows."

"Definitely." Rebecca sipped her tea. "They're so cute!"

"And I need to give them a rewarding social life."

Rebecca didn't notice the sarcasm. "Not really. If you get two at once, they should be happy together."

"Besides," Peter reasoned, "you can make money from cows. If you raise them up and get them pregnant, you can get like twelve hundred dollars for one cow."

"But what about when Katya and Derek come back?" I said, starting to sound a little desperate. "We're gonna have a pregnant goat, and this maniac rooster, and now you want cows—what are we gonna do with the animals?"

"Exactly." Peter grinned. "Now we can't move back to the city."

So the next morning, I strode across the road hoping to glean some wisdom from Hamish. I found him in his milking shed, cleaning out the machinery.

"We're thinking of getting calves," I announced cheerfully.

Hamish straightened, keeping me in the crosshairs of his skeptical gaze. I touched the top of my head to make sure I wasn't wearing my devil's horns.

"Ya don't want calves," he said finally. He sounded grim, and I wondered for the umpteenth time if he hated me.

"I don't?"

"Nah. Too hard."

"But I *like* a challenge," I protested. "And my niece thinks they're cute."

This was possibly not the best thing to say to an overtired dairy farmer. Hamish looked at me. "If you just want to bottle-feed some calves," he suggested, "come over to the calf sheds anytime. I'll let you feed the young ones."

That was a little frustrating. Hamish thought I was not only a city slicker, but also an idiot. I didn't want to pet his cows. I wanted to keep my own for milk.

"I want to make cheese," I floundered. But he was already shooting high-pressure water through his milking machine, and the conversation had come to an end.

What did I expect? I thought miserably. *He probably saw me chasing after a rogue cow while bleating like a sheep. He comes around to borrow a lamb teat, and we're giving hand jobs to alpacas.*

But I am a stubborn fool, and there's nothing like getting dissed by a farmer to urge me ahead. Rebecca knew a lot about cows, but she'd always worked under the supervision of a teacher. This time, we'd be in charge. So I went back to the library.

Moving through the agricultural aisles, I cleared the shelves of all books having to do with calf raising. Then I wrote out a list of questions and made an appointment with our veterinarian. At the end of my hour-long visit, I knew quite a lot about cows: types of feed, vaccinations, and something called "drenching," which is when you shoot medicine down the calf's throat to kill all the worms in its gut. Our veterinarian must have enjoyed my zeal, because after my visit she wrote up a four-page report on calf care and sent it out to us in the mail.

"See?" I crooned, waggling the calf report under Peter's nose. "Grumpy farmers aren't the only ones who know something about cows."

Armed with our new information, we started scanning the listings. Buying calves, we discovered, was easy. We found a farmer on the Internet who had some Jersey calves for sale, rang him up, and struck a deal on the spot. At this stage, I was feeling pretty confident about my farming skills—so bold that I decided to bring the calves home in the back of my station wagon. This seemed like a sensible idea. There was plenty of room back there, and three-day-old calves aren't much bigger than a couple of dogs.

I even asked our vet about it first. "Shouldn't be a problem," she told me. "There's some risk of scours, but it won't hurt *them* to ride in the back of a car. All you need is an electrolyte solution to rehydrate them, and they'll be good as gold."

In retrospect, I should have looked up the word *scours* in the dictionary. Like *VX* and *sarin*, the word sounds deceptively benign, like it might have something to do with cleaning: *She scoured that pot till it glowed. He scoured his face with hot soap and water.* In calves, however, scours refers to a death slurry that comes shooting out their assholes. I suppose you might call it diarrhea, but that would be missing the point. The stench of calf scours is so obscene that you cannot inhale it without gagging. It's like getting bombed with mustard gas or one of those other horrors long outlawed by the Geneva Protocol. Scours, in short, are a war crime.

But the vet didn't mention war crimes, so we duct-taped some feed bags to the floor of the station wagon's trunk and thought that would suffice. And at first, everything seemed to go fine. For starters, Rebecca was right. Baby calves are outrageously cute. We chose two caramel-colored Jerseys with huge dark eyes the size of salad plates. When they lowered their eyelashes and flicked their big soft ears, I felt like I'd teleported into a Christian greeting card.

Peter, Rebecca, and I loaded them into the car and set off for home, with Peter at the wheel. We took the back roads alongside pleasant green paddocks, and the calves seemed to enjoy the scenery.

Then they stuck out their tongues. "Mama!" I heard from the backseat. "The cow is *licking* me!" I turned to see a python-like tongue emerging from a calf's mouth as it strained to suck my daughter's hair. Rebecca, who was wedged between the two car seats in back, was crouched down low to avoid the tongue. She appeared to be busting up laughing.

"Home! Home!" Silas insisted, batting at the tongue snake on his head. For such adorable animals, these calves had unbelievable tongues. They were huge, black, and pointy at the end, like Satan's.

Just then, Silas's arm connected with a calf tongue. Maybe he swatted a tad too hard, but the next thing I heard was something wet hit the feed bags, and the world became a blur.

"Sugar, Honey, Ice, and Tea!" Rebecca yelled the phrase she often substituted for expletives when in the presence of children. It was cute, but I only dimly registered what she was saying.

"*Air...*" I gasped, rolling down the passenger-side window. I arched my neck out of the car, fully prepared to vomit on the tarmac. Beside me, in the driver's seat, Peter was coughing and gagging. He swerved to the wrong side of the road, his hands going into a spasm.

"*Mama*," Miranda wanted to know. "Did you *fart*?"

But I couldn't speak. The world was swimming. Another audible squirt came from the back, and the air in the car turned green.

Then they started slipping in it. Having thoroughly fouled the feed bags, the calves began sliding back and forth in their own filth. One of them fell, whipping up a new cloud of stench. The other one collapsed on the first, and they both emptied their bowels in unison.

I have no memory of the remainder of the trip. The next thing I remember is hosing down the calves in our driveway and drying them off with towels. The dogs took one sniff and ran away, cowering at the far side of the house. I ran inside to mix an electrolyte solution, hoping to cure the calves of the scours.

Once they'd slurped down the bright yellow drink, we sat in the paddock and tried to choose names. Mustard and Chlorine were vetoed for being too depressing, as were VX and Sarin. Rebecca came up with Luna and Maya, but I reminded her that these were cows and not backup singers for the Jerry Garcia Band. Then I suggested Plague and Ebola, but Peter's no fun and declined.

"It has to be cuter," he insisted. "Like Daisy and Buttercup."

"My cow at Putney was named Belinda," Rebecca mused wistfully. "I called her Bebe. She was such a beautiful girl. I used to weave flower garlands and drape them around her neck."

"Really?" I rolled my eyes. "When I went to school, I used to read books."

Rebecca looked a little hurt. "Hey, now," Peter nudged me. "Not everyone was a front-row nerd."

"You're right," I conceded. "Sorry. I probably should have spent more time with flowers."

"That lil' lady looks like cinnamon," Miranda said, pointing to the darker of the two calves.

"That's not bad," I commented. "Cinnamon. Maybe we'll call her Sin for short."

"And Lil' Lady?" Miranda asked.

"Cinnamon and Lil' Lady?" All I could think was *Now my cows have stripper names*. But with my outside voice, I said, "Great choice, Miranda. Those are perfect."

We sat with this for a moment, watching the calves. Cinnamon extended her snake's tongue and twisted it upward, delicately picking her own nose. This was an impressive party trick, and I thought to myself, *Yes. Stripper names will do just fine.*

CHAPTER SEVEN

BACK FROM LOVE MOUNTAIN

Unfortunately, the calf incident was not our last brush with the scours. While Peter was at work during the week, Rebecca and I spent our afternoons together. I'd emerge from my office around noon, and then we'd make lunch and bake some bread, and Rebecca might get out her spinning wheel before we left to pick up the kids.

Our afternoons were mostly uneventful, until the Day of the Indian Chili Bomb. Looking back, I have to admit I had no one to blame but myself. After my and Peter's trip through Mexico and Central America, I'd come away with a monkey on my back: a devastating addiction to hot and spicy chilies.

These were difficult to find in New Zealand. Most of the time, I'd bring home a jar of jalapeños only to discover with a sinking heart that the spice had been boiled right out of them, their flavor sanitized with sugar and brine. Tongue-numbing habañeros simply weren't on the menu for the majority of Kiwis, and though exotic chilies could be purchased in the major cities, Whangarei was a wasteland for spice.

So when Rebecca suggested making tacos for lunch, I pulled out a new jar I'd found, full of green and red chilies that had flown all the way from India. Their South Asian origin was promising, and I placed one on my tongue, heart brimming with hope.

Instantly, my cheeks flushed red and my eyes began to water. The balance of sour, salt, and spice was just what I needed to scratch that hot-pepper itch. Rebecca piled our plates with tortillas, rice, and beans, and when I gave her a chili, she reached for her water glass.

"Wow," she commented. "These are pretty strong. One's probably enough for me."

Which was fine. We all have our limits. As for me, I ate seventeen. I piled them on my beans and I stacked them in my taco. Then I started wrapping them in slices of ham, merrily calling them chili bombs and popping them in my mouth like candy.

After our meal, Rebecca dragged out her spinning wheel and started feeding it a pile of alpaca fleece she'd brought with her, calmly pressing the foot pedal to get the little wheel to spin. I was still planning to make a pashmina, so she'd started me on the lowest rung of the fiber ladder: picking fleece.

This is a mind-numbing task in which you sit like a mental patient, carefully picking tiny flecks of dirt out of endless fluff balls. I suppose with the right mind-set, such as cocaine psychosis, this would be fascinating, but after five minutes, my brain went numb and I turned on my iPod. I started up the audio version of *Middlemarch*, reasoning that this way I might actually finish that eight-hundred-page nightmare.

About an hour passed, and Dorothea Brooke was just starting to crush hard on Edward Casaubon, when I felt a peculiar tingling in my abdomen. *That's funny*, I thought. *It's not that time of the month yet.* I shifted my weight on the couch and turned my attention back to my fluff.

A little later I checked my watch and stood up, brushing stray fur balls off the front of my jeans. "We'd better get going," I told Rebecca. "It's almost time to get Miranda from preschool."

"Oh, sorry, sure!" Rebecca pushed her spinning wheel to the side and hurried to her feet. Although Becca was technically my niece, I hardly knew her. Peter and I had spent most of the past seven years abroad, and though we'd had some visits over the years, this girl was practically a stranger to me. We were always friendly with each other, but a veneer of good manners kept our conversations stiff.

I grabbed my purse, pulled out the car, and Rebecca climbed in. She rifled through our CD collection, a motley pile of generic-looking disks burned from iTunes. "What's this?" she asked, holding up one of the few with a commercial label.

"CMCB?" I grinned, turning back to the road. "Chinese MC Brothers. They're rappers from China. They're pretty good, actually. Check it out."

I'd always been fond of this CD. Blending classical piano, hip-hop and metal, these guys could really crank a tune, and everything sounded extra angry in Mandarin. Rebecca pushed the volume up to window-thrumming levels.

At about the time we pulled onto the main road into town, my cramps returned. And now their origin was clear. They had nothing to do with my period; rather, it was the seventeen Indian chilies I'd popped with my lunch. My lower intestine was starting to scream.

But "just a moment, I've got the shits" is not a phrase I felt I could say to a delicate young woman I hardly knew, so I determined to wait it out. We were twenty-five minutes from the preschool, I reasoned. I could do anything for twenty-five minutes. *This is a breeze,* I told myself as cold sweat beaded my brow.

CMCB launched into one of their metal numbers, and as the lush green trees of Purua sped past on both sides, the bass chords throbbed in my skull. "*Qī dài nà yí kè,*" the lead vocalist bellowed, and each beat made my sphincter clench tighter. On the steering wheel, my knuckles went white.

Rebecca, for her part, just thought I was dancing, and she started bouncing up and down in the passenger seat, doing a free-spirit version of busting a move. I could feel my colon contracting, an invisible python recoiling for the strike. I drove stiff-backed, clenching the steering wheel, rocking my body back and forth in a paroxysm of pain.

By this time, we'd left Purua and were rolling past the manicured lawns of Ruatangata. These were the moneyed suburbs, the place where city-based professionals kept a few rolling acres, two luxury cars, and a swimming pool. Unfortunately for me, what they conspicuously did not keep were public toilets.

I'm not saying I'm special. We've all been in uncomfortable bathroom situations before. I've been stuck on a Guatemalan bus with a violent stomach bug, I've suppressed my bodily urges in classrooms and on job interviews. I gave birth to both my children naturally, each time powering through waves of contractions that threatened to rip my back in two.

Which was why I knew when I'd had enough. As with childbirth, there comes a time when you just don't give a fuck. "There's something I need to tell you," I gasped, not daring to glance at Rebecca. "I am about to have violent diarrhea."

To my surprise, she didn't squeal. Rebecca might be a chamomile-drinking tree hugger, but she's also a farm girl, and she knew just what to do. "Here, up here." She pointed. "There's space to pull over. Go behind that tree. There's no one around."

With a buoyant sense of release from the darkest void of hell, I pulled the car off the road and slowed to a stop. Already loosening my belt buckle, I lunged for the door handle, stumbled onto the gravel, and ran toward the nearest property.

"Wait!" Rebecca called. "Do you need Kleenex?"

She handed me a tissue, and I tumbled into a stand of trees, thanking God, heaven, and all my lucky stars that the good people of Ruatangata worked for a living. No one would be here to witness my disgrace. I crouched down and let fly, a fusillade of South Asian shrapnel defiling the grass beneath me.

I straightened, my body limp with relief, and froze. Straight ahead, not twenty feet away, was a comfortable home with a large picture window. The back of a gray leather couch was visible, and the dim, shifting glow of a television set. And there, like the all-seeing guard of the Panopticon, was the back of a woman's head. She was watching TV.

With the smooth, silent moves of a poop ninja, I wiped myself off and reached down for my pants. The gray head tilted, leaned forward. I stopped. *Is she turning?* What would she do, I wondered, if she turned around only to find a half-naked woman in her carefully groomed backyard, pants around her ankles in a puddle of filth?

I fastened my trousers. Silently I backed away.

"All better?" Rebecca asked when I'd safely returned to the car.

"All better." I nodded. And from then on, the two of us were the best of friends.

Besides allowing me to bond with my niece, the Day of the Indian Chili Bomb helped me to empathize with Cinnamon and Lil' Lady. Everyone has a bathroom situation now and then, and those poor cows didn't have the choice to pull over to a nice green lawn. Once they got over their scours, the girls were much more appealing. We'd go out and visit them in the evenings, scratching them behind the ears and brushing dried dung from their caramel coats. Occasionally, Cinnamon regaled us with her fantastic nose-picking trick. But even though we now had two real farm animals, we still weren't winning any points with Hamish. This is because we asked him if we could feed his leftover milk to our cows.

I know cows are supposed to give milk, but for the first few months, they just drink it. Powdered milk is expensive, and not as nutritious for the calves. Because we lived across from a dairy farm, we asked Hamish for a bucket of milk each day. He was kind enough to agree, and he didn't even charge us, which made me think he might be a really nice guy under that constant frown.

The problem was one of perspective. From Hamish's standpoint, cows were livestock. They needed a certain amount of protein and calories each day to survive. To this end, he gave us curdled milk, milk with antibiotics in it, and sometimes there were leaves and twigs floating on top.

I found this unappealing. Cinnamon and Lil' Lady were pets, and we envisaged their mealtimes as a chance for a pleasant gastronomic experience. I'd pull my gumboots on over my pajamas, put on a pair of rabbit ears to keep my hair out of my face, and fill two large calf feeders with fresh, warm milk. Sometimes I'd beat a raw egg in there for extra vitamins. Then I'd strut out to the calf pen with the feeders clutched to my chest, the rubber teats protruding like bright pink carnival nipples, and sing little songs to the calves while they ate. They were particularly fond of "You Are My Sunshine," which I sang out of tune while giving them a good scratch behind the ears.

Hamish thought this was ridiculous. "You don't need to *warm* the milk," he told Peter one day when he was filling our bucket. "They'll drink it any which way they get it."

"Could I have different milk?" Peter asked, peering skeptically at the bucket. "This one has a snail floating in it."

"Give it here." Hamish took the bucket and flicked the snail out with a grubby thumb. He handed it back. "See? No snail."

When Peter reported this exchange, I considered that Hamish and I would never see eye to eye on the nuances of a gourmet calf diet, so eventually I stopped asking for free milk. Instead, I went back to my calf book.

"Bad news," I announced one night over dinner.

"What, Mama? What happened?" Miranda asked through a mouthful of pasta. "Did more houses burn down?"

"We're not getting milk for two years."

Rebecca lifted a napkin to her lips, delicately suppressing a smile.

"Oh, for Christ sake." Peter set down his fork. "Why not?"

Miranda looked impressed. "Is *Christ* a grown-up word?"

I ignored her. "Because the cows have to have babies to make milk, duh. And we can't breed them for a year. Then it takes another year to gestate a new calf."

Peter was crestfallen. In retrospect, this salient point of cow biology was obvious, but it came as a shock to us at the time. I thought cows made milk like chickens made eggs. I imagined a spigot somewhere on the bottom, like the lever you press for your Coke at McDonald's. I never connected cow's milk with sex and lactation, despite having breast-fed two babies of my own.

"But goats are much faster," I told Peter, trying to cheer him up. "And Pearl's had two cycles on Love Mountain. We can pick her up soon." I checked the calendar. "We can even go tomorrow, if you want."

Peter perked up. "We can?" He immediately pushed back his chair and called Jackie, the neighbor who owned Love Mountain. "It's Peter," he said when she answered. "My goat Pearl's on your hill up there, and I was wondering if we could get the key to the gate."

Becca and I watched his face as his eager grin faded and his brow creased. "Uh-huh," he said. "Okay. We'll be up there tomorrow."

He hung up the phone. "What's wrong?" Rebecca asked. "Is Pearl okay?"

"She's fine." Peter hesitated. "I think she's fine. Jackie said she saw her today."

"Well, that's great!" I enthused. "We won't have any trouble finding her!"

"It's not that." Peter came back to the table and sat down. "She said she saw 'a pile of boys, all clustered in a heap. And your girl was right there in the middle.'"

"So, like, a big pile of male goats?" Rebecca looked confused. "On top of Pearl?"

Peter nodded.

"Oh. Wow."

That phone conversation was a wake-up call. All those weeks she was up on Love Mountain, we'd managed to ignore the reality of goat-mating rituals. Instead we imagined Pearl frolicking in the sunshine, snacking on daisies and grass. We'd somehow even convinced ourselves the bucks would line up and take turns. Now terrible images flashed through my mind. *How many bucks in rut? How many gallons of pee?* I shuddered.

The next morning, Rebecca and I set off to rescue our goat. It wasn't hard to find her among all the wild bucks, because Pearl was the only one with a collar. We offered her bananas, but she wasn't interested. We ended up herding her back to the car.

That night, Peter looked worried. "She seems skinny," he observed.

"She seems depressed," I said. "Do you think she resents us?"

Peter made a face. "Stop anthropomorphizing the goat. She lived on a hill for a month and got laid. It's goat paradise. She's fine."

But Pearl seemed listless. She ate a little grass and drank some water, then sat down under the palm tree and rested. Before I went to bed I stroked her back, right along the black stripe on her spine where she liked it best. She bleated softly, butting my hand with her head. I could feel her bones through the fur.

The next morning, I looked out the window after finishing my coffee. And there was Kowhai, barking and prancing over a very limp goat carcass. "Kowhai!" I shrieked. *"Baaaaa!"*

Kowhai looked up, as surprised as I was that Pearl had fallen. The game was supposed to go like this: Kowhai tried to bite Pearl's legs, and Pearl reared up until Kowhai ran away. But this time, Pearl hadn't fought back. She'd just collapsed.

I rushed to Pearl's side, quickly noting that she was still alive. She struggled to her feet, so she had no broken bones, and I couldn't find any blood. But then I had to look for the goat abortions.

Our farming books had warned that stress and traumatic experiences could cause spontaneous abortion in a pregnant doe, and they urged me to keep our goat safe and comfortable during gestation. *Does that include the gang rape?* I wondered. Because now that we'd done it, the process of getting a goat pregnant in the first place seemed about as traumatic as things could get. After that, a couple of dog bites would feel like a holiday.

Still, I worried about the possibility of miscarriage. For the rest of the day, I followed Pearl around, cringing every time she raised her tail. On a goat, all the sex gear is arranged close together in the hind quarters, so when I saw her tense up, I'd brace myself, waiting for baby bucklings to shoot out like paint balls. But instead, it would just be pee or a few dozen goat berries, and soon I stopped fretting, convinced her baby was okay.

Pearl, however, was another matter. Her lack of fighting spirit concerned me. I wasn't about to lose another animal. So that afternoon, I rang up the veterinary office.

"You'll have to bring in a stool sample," the receptionist chimed. "We can't tell what it is without a sample. And not a little teaspoon, either. We'll be wanting a nice big soupspoon full of goat berries!"

"Okay." I nodded grimly. "You got it."

It wasn't hard to collect the poop, since Pearl was basically a grass-to-poop conversion machine, but it was tough making time in my schedule to get to the veterinary office. This meant that for the next several days, I had a mustard jar full of goat turds in the bottom of my handbag.

The poop in my purse didn't really bother me as much as the logistics of the thing. As a busy mother of two small children, I often found myself rooting blindly in the depths of that bag. A mustard jar full of goat berries could feel very much like the smooth surface of a water bottle, for example, and when I offered it to my thirsty daughter on the way to the vet's to drop it off, her eyes grew wide with joy. "Mama," she asked incredulously, "is that *candy*?"

"No," I snapped, shoving the jar back in my bag. "It's *poop* and it's *poison* and also very *spicy*. So don't eat it."

"But it *looks* like candy," she protested.

I didn't answer. Mutely, I floored it to the vet's office.

Once we opened the tinkling door to the waiting room and addressed ourselves to the cheerful receptionist, everything began to feel a bit more normal. "I have, er, something for you," I told the lady behind the counter, pulling out my mustard jar. The receptionist glanced at the label:

> FAECAL SAMPLE FOR "PEARL"
> (GOAT) SYMPTOMS: WEIGHT LOSS,
> LETHARGY, PALE GUMS

I'd labored over that label. I mean, really stressed about it. The extra *a* in "faecal," for example, just looked pretentious. But here we were living in New Zealand, the land of hobbits, faeces, and diarrhoea, so I was trying to spell like a local. Then why had I specified the species? This was a veterinary office. Did I really need to tell them it was a jarful of goat shit? Was there the slightest chance that, like my daughter, they would mistake it for a selection of tiny truffles? Hand rolled, perhaps, and packaged in a pretty Dijon mustard jar? Finally, why had I written down the symptoms? The lab didn't care. They were just a bunch of underpaid technicians who spent all day hunting goat poop for worm eggs.

No, I think I wanted to make something clear: that I am not a person who fills jars with goat turds for fun. That this was not a crude practical joke, but rather a serious attempt to cure my goat, who was clearly ill. It was probably a parasite infection, unless she'd eaten a poisonous plant. Or maybe our dog had injured her, unless she was just depressed because we'd left her on a hillside to get gang-banged for six weeks.

So when I spoke to the vet, I tried to play it cool. "My goat is, uh, thin," I began, racking my brain for technical phrases.

"Male or female?" the vet wanted to know.

"She's a . . . ewe." *No, shit, that's a sheep.* "I mean a girl. I mean a doe."

The veterinarian jotted this down on her index card. "Do you know her weight, by any chance?"

Now there was a stumper. Because how do you weigh a goat? Was I supposed to coax her onto a bathroom scale? Do I pick her up and weigh her myself? This animal had hooves, and teeth like a pearly white chainsaw. If she didn't kick me in the gut, she would probably eat my face.

I took a shot in the dark. "Um . . . a hundred pounds?"

"It's most probably worms," the vet informed me. "*Haemonchus contortus.* We call it the barber pole worm. We'll have to get right on top of it, so we'll give you a prescription today. They suck the goat's blood, you see."

I swallowed. "They do?"

"And they reproduce rather quickly. Up to ten thousand eggs a day. So the worms can suck a lot of blood in a very short time."

The fact that vampire worms were reproducing by the tens of thousands in the belly of my goat should not have come as a surprise. By now I'd learned that country life is not a pastoral painting. Sure, at various times during the year you might see fluffy white lambs prancing in the tall grass, but those moments are rare. Real country life, it turns out, involves blood, shit, and worms. Also goat abortions.

The other thing it involves is dead chickens. I found that out when I got home from the vet's office only to learn that Kowhai had chomped on a chicken.

Through all our recent animal traumas—the brain-eating alpacas, the scouring calves, the worm-ridden goat—we'd kept our flock of six chickens, along with the notorious Jabberwocky. They all appeared healthy enough. There weren't any more leg mites, no one was showing signs of leprosy or cloacal trauma, and they were laying a regular supply of eggs. Their biggest problem was the rooster and his insatiable sex drive.

That is, until Miranda left the door of the chicken coop open. At first I didn't know anything about it. I dosed Pearl with the worm medicine, and I was just taking my coat off when Miranda came running. "Mama!" she hollered, naked apart from purple underwear and her shiny black gumboots. "Kowhai did bite a chicken!"

Oh God, here we go, I thought, running after her. *My daughter's first taste of mortality.* After all, we'd executed chickens and roasted a sheep on the property, but she'd never actually witnessed the carnage. Would she be traumatized? Have nightmares? "Is it Jabberwocky?" I asked hopefully.

"No, it's just a chicken," she said, leading me to a large flax bush by the henhouse, where Kowhai was happily chewing on a pile of feathers.

"*Kowhai!*" I shouted in my Wrath of a Vengeful God voice. "*Baaaaaa!*"

Kowhai dropped the chicken, her eyes mourning the wasted snack.

My medical diagnostic skills were not what they could have been. "Chicken?" I asked. "Are you okay?"

Miranda poked the bird with a paint-stained finger. "I think it's dead, Mama. I think it's dead."

I had to agree. The chicken was looking decidedly not okay, lying limp on the ground, her eyes all white.

But she wasn't dead, just in shock. Those white things must have been chicken eyelids, because in a minute or two her eyes opened wide and she stumbled wearily to her feet.

"It's not dead, Mama!" Miranda crowed. "I think the chicken's not dead!"

"God, I hope not," I agreed, picturing the Chicken from Beyond the Grave. "Because that would be weird."

That night, over dinner, I had a new discovery to announce. "Her wing is broken," I reported. "There's no blood, but the left wing is hanging down, and I think it hurts."

Peter looked at me blandly. "Should we stick it in the freezer?"

"Uncle Peter!" Rebecca stopped chewing her veggie burger to fix him with an indignant stare.

"*No.*" I shook my head, vehemently. "We have to bind her wings. I read it in my chicken book. We bind her wings with gauze, then just keep her away from the rest of the flock."

"How do we do that?" Peter asked, cutting into his chicken. The one on his plate was sautéed in a nice, creamy white wine sauce.

"We'll keep her in a cage. On the deck."

"Or else what?"

"Or the other chickens will torture her. You know, like the pecking order."

Peter waggled a forkful of chicken at me. "Tell me again. Why don't I just cut its head off and stick it in the freezer?" Perceiving that I was on the side of the chicken, Rebecca kept her arms crossed and listened intently.

"Just because she's *injured*? You want to *decapitate* her? Because she's not *perfect*?"

"Because she's food."

"*I'm* not perfect. My thighs have cellulite. Are you gonna cut my head off, too?"

Peter reached over and patted my knee. "Don't be silly, honey. You don't have *that* much cellulite." Surprisingly, this didn't make me feel better.

Peter wanted nothing to do with the chicken physiotherapy

project. So Rebecca and I dug up some gauze from our first aid kit, then she held the bird firmly on her lap while I bound her wings. We made a little house for her in a large dog cage on the deck, which quickly became known as the Chicken Hospital. She looked depressed in there, but she ate and drank, and she didn't seem to be dying. Every few days, we'd check her dressing, and I'll be damned if her wings didn't start looking a lot more even.

Pearl was getting better, too. Once we gave her the worm medicine, she started gaining weight and defending her territory against Kowhai. After a while, people started admiring her. Autumn and Patrice came around for a visit, and she stroked Pearl's ears while he examined her udders.

A handsome man with a strong Gallic nose and bright blue eyes, Patrice had spent years abusing himself as a chef in top French restaurants before burning out and moving to New Zealand. Now he worked as a teacher's aide, helping Silas through the routines of a school day: hanging up his bag, sitting at a desk, and beginning to write with a pen. At first I thought Patrice was naturally quiet, but then I realized he struggled with English, so sometimes he gave up on talking and kept his thoughts to himself. This time, however, he couldn't hold back.

"You are so lucky," he told me enviously. "Your goat has great tits." I wasn't sure what he meant by this until I realized it was the French accent that was confusing. What he intended to say was "teats," and his interest in Pearl extended just to chèvre and fresh milk.

The funny thing is, he was right. Pearl *did* have great tits. The more I learned about dairy goats, the more I learned about teat position, and how you want them angled the right way to make milking easier. And once we killed off the vampire worms and once the troubling memories of Love Mountain were behind us, I began to feel real affection for Pearl. I watched her belly growing each day, murmuring sweet nothings to her and scratching her between the ears. As for Rebecca, she made gourmet goat salads for Pearl, pinching off the tenderest shoots of bok choi and chard, then sprinkling the dish with nasturtium petals.

As far as we could calculate, Pearl would be due sometime in November, a mere two months before the end of our rental. We had to find a safe place for her baby, but week after week, nothing was for sale but multimillion-dollar farms—and the illegal Slovak house, with no heat or phone service. Despite her gourmet goat salads, it was starting to look as if Pearl might be raising her kid in a car.

CHAPTER EIGHT

TURKEY TIME

Then, at the end of June, another opportunity appeared. It was Monday afternoon and Amanda was over for a visit with her kids. We poured drinks, and Amanda pushed Amelia on the giant tree swing that hung from the ancient totara on the property. She could push Amelia with only one hand because she was balancing a glass of quince wine in the other.

"I've been thinking," she mused. "What about Michiko's house?"

"The burned one?" I frowned. "I was sort of thinking something with a roof."

"You'd be helping her out. There's no way she can live there. Then you could build a new house, maybe start building one now, while you're here."

"So you've been in touch with her? Since the fire?"

Amanda shook her head. "No. I don't know where she's staying, but I think you could make contact through her lawyer."

I hadn't thought of building a new home. And I didn't want to swoop in on Michiko's tragedy like some tacky American jackal. But if buying the land might help her out, then that was something else.

Amanda scribbled out directions to the house on the back of a grocery receipt, and I stuck it to the refrigerator with a magnet. For the rest of the week, Peter and I talked about this new possibility. We pored over house building websites, considering that we might get much better value if we bought cheap land and built a new home on it. Plus, Michiko's property was right up the road from the school. The location was perfect.

On Saturday, we decided to drive there. "I don't know if I want the kids wandering around a burned-out ruin," I said, glancing at Silas, who was pensively tasting the corner of a used Kleenex.

"That's okay," Rebecca piped up. "I have some fabric dyeing I want to work on. I'll take care of them."

"But I *want* to come!" Miranda howled, her eyes brimming with tears.

"Miranda," Rebecca beckoned mysteriously. "We're going to make a *magic potion* this morning, and I need your help."

Miranda considered this option. "Magic?" she asked doubtfully.

"Yes! With onions!"

Mollified, Miranda agreed to stay.

"That girl is incredible with the kids," I commented, once Peter and I were pulling out onto the road. "How does she do it?"

"Becca's just talented," Peter said, shrugging. "And then there's Will. She's always been protective of him." Will was Rebecca's little brother. Five years younger, he had been born with a developmental delay. No one was sure exactly what was wrong, but he'd always been different, and from the start, Rebecca knew how to love someone who was vulnerable.

Michiko's property was situated on a loose gravel road, deep in the Purua countryside. We began to see bright yellow signs nailed to the trunks of overgrown trees. "Dogs Kill Kiwi," they adjured, and I swallowed hard. There's a hundred-thousand-dollar fine if your dog kills a kiwi bird in New Zealand. Living out here with Kowhai could get expensive.

Peter took notice. "We're right next to the reserve." He looked around, casting a sharp eye toward the bushes at the side of the road. "I bet you really can see wild kiwi out here."

After three or four minutes, we reached the crest of the hill. Peter glanced at the address on the small slip of paper. "This is it," he said. "You'll have to open the gate."

I climbed out of the car and stopped, taking in the view. The land was perched on the edge of a hillside, looking out on a deep gorge full of native New Zealand bush. Manuka, kanuka, and giant ponga ferns were woven together in a rich prehistoric foliage. Shafts of sunlight shone down from pink clouds, bathing the valley in golden light.

I opened the heavy steel gate and followed our station wagon down the driveway. Here were the ruins of Michiko's house. Sturdy brick walls were now piles of debris, the site a foot deep in ground glass and twisted metal. A single apple tree stood on the pavement, its branches singed.

We picked our way through the rubble, careful not to cut ourselves. Personal items were scattered through the wreck: a splintered skateboard, a little green flip-flop. A few pages of sheet music floated through the ashes. There was a Chopin prelude and "The Girl from Ipanema." I could imagine Michiko playing that one for her kids. But it felt too intimate standing in the ruined lounge, as though we were trespassing in Michiko's home. After a quick look around, we made our way back outside.

"The concrete pad might be still good," Peter commented, kicking it thoughtfully with his boot. "If we could clear the bricks out and keep the foundation, we'd save a ton of money."

The property was positioned on a level building site carved from the hillside. On three sides, the land dropped away, quickly surrendering to thick brush and wild foliage. We started picking a path down the slope, dodging thorny bushes studded with bright yellow flowers.

Peter pinched off a yellow blossom and crushed it between his thumb and forefinger, sniffing the petals. "This is gorse, I think. Scottish gorse. Smell the coconut? This stuff's a nightmare to get rid of."

"Well, we'd have to do something about it." A thorny branch tugged at my coat sleeve, and I yanked it back before the fabric could tear. "It's growing everywhere."

At the base of the gorge we found a shaggy macrocarpa tree hanging low over a cool, dark stream. Silty with rainfall, the water flowed thick like melted chocolate.

"Be a good place to read in the summer," I pointed out.

Peter reached down and wet his hands in the stream. "I could make a little bench here, maybe put in a fountain."

I smiled at the thought and looked up. "Who's that?"

A shadowy figure stood at the top of the hill. I couldn't make him out at this distance, but he was working the latch on the gate. He unlocked the barrier, swung the gate wide, and came onto the property.

"That's not the husband, is it?"

Peter glanced up sharply, shading his eyes with his hand. "Come on." He took my hand protectively and started leading me back up the hill.

Whoever he was, the man had spotted us. He'd left the gate open and was now sauntering down the driveway, calm as can be. He looked casual, like he owned the place.

"Do you know what Duncan looks like?"

I shook my head.

"Me, neither." Peter's face tightened. "Let's see what he wants."

We scrambled up the hill, my boots losing purchase on the crumbling soil. Peter grabbed my arm as I started to slip. The man was staring in our direction. The sun shone behind him, casting his face in shadow. He wore a long dark coat and what looked like ragged jeans.

Twenty feet away, he called out. "Howzit guys? All good?"

"Oh, God." I let out my breath. "It's Skin."

We came up the last rise of the hill. Skin stepped back, a mirthful smile on his craggy face. He turned to Peter. "Ya gonna buy this place, then? House needs a bit of work."

"I don't know. What do you think?"

Skin's eyes shone. "Good land. Got lots of pretty flowers on it."

"You mean the gorse?" Peter grinned. "This stuff is the hedge from hell, isn't it?"

"Nah, she'll be right. Just put a whole heap of goats on it. They'll get rid of it for ya." Skin stood there, hands shoved in the pockets of his jeans, surveying the view.

I followed his gaze, out over the valley full of lush, green bush. Most of it was the hedge from hell. But it would still be dark out there at night, an ocean of trees surrounding us. Not a boat maybe, but still a new horizon.

"Yeah," I said. "Goats. I could see goats on this land." And secretly, I thought, *I don't care if it's not perfect. I could see us here, too.*

We drove back home then, and when we walked in the door, we

found all our windows fogged up with steam. A cauldron of onion skins bubbled on the stove, and Rebecca was happily curled up on the couch sewing pistachio shells into a nightgown.

"There's a reason for this, right?" I asked. "Besides the onset of schizophrenia?"

"Becca's dyeing!" Miranda informed me. "And the 'stachios are the zist!"

A few extra shells tumbled to the ground. Silas placed these into his mouth, licking the salt from each one and then handing them back to Rebecca.

"*Resist*," Rebecca corrected Miranda. "It's a technique my fibers teacher taught me. We stitch the nuts into the fabric. Then, when we dye it, we get this really cool pattern. I want to try it with avocados, too. And lichen."

Dyeing was an element of fiber art that I hadn't considered when we first asked Rebecca to join us. I knew she was into weaving and knitting, but natural dyes could be harder to live with. Our countertop was cluttered with empty ice cream boxes where we sorted our kitchen scraps, the first three labeled "Compost," "Chickens," and "Goat." Now we had a fourth vessel, this one titled "Becca Box." The Becca Box contained decaying food items that might serve as experimental dye ingredients: moldering chunks of red cabbage and chard, puckered avocado skins, and piles of furry gray lichen. The whole assortment was starting to stink, and I was running out of counter space.

"So I was thinking," Rebecca inquired. "What are we doing for Thanksgiving?"

Peter and I exchanged a look. Apart from a very sandy turkey we'd roasted on a Costa Rican beach a few years ago, we'd mostly let the holiday slide. Turkey could be hard to find outside

North America. Since it was so often dry and overcooked, the people we'd met on our travels tended to consider it a second-class meat.

"Hadn't really thought about it," Peter admitted, pulling a cold bottle of water from the fridge. "Seeing as how it's June."

"But that's the thing." Becca put down her nut-nighty project and looked up. "We slaughtered them when they were six months old back at Putney. So if we get some young turkeys now, we could serve our own birds for Thanksgiving."

"Like baby turkey chicks?" I wondered if they cut off their noses.

Rebecca shook her head. "Not the babies. I don't want to sound mean, but baby turkeys are as dumb as they come. At Putney, they were always drowning themselves in their water dispenser. When they weren't lighting themselves on fire."

I blinked. "Why were they lighting themselves on fire?"

"It wasn't on purpose. They were just so effing *dumb*, so they'd wander into the heat lamp and set their fluff on fire."

"Like Michael Jackson," Peter commented.

"Yeah." Becca smiled. "Kind of. No, I was thinking we could get some young ones and, you know, raise them up."

"I thought you were a vegetarian?" I queried.

"Mostly." Rebecca picked up a pistachio shell and resumed stitching. "But I'll taste an animal if it had a happy life. You know, like happy meat."

"Happy meat," Peter repeated.

"Yeah."

To prepare for our happy young turkeys, Becca and I determined to clean out a small pen that was already built beside the storage sheds. About forty feet square, it contained a hovel that would provide shelter from the weather, and the wooden enclosure seemed sturdy enough. The problem was that it contained a rainforest.

Katya and Derek hadn't used this pen for some time, and in the intervening months, it had become choked with weeds. These were not small clusters of dandelions. The pen was bursting with formidable plants, many of them over six feet tall, green and shaggy with long, pointy leaves and clusters of dark berries.

Pulling on our gumboots, we got to work. As tall as they were, the weeds came up easily, and we lugged armloads of them to the wheelbarrow, taking turns carting them to the palm tree for Pearl to snack on. Though the day was cool, it wasn't long before we were both dripping with sweat. Rebecca had her iPod connected to some speakers, and we kept up our energy by blasting Reggaeton and New York hip-hop at obnoxious levels. Which is why we didn't hear Amanda drive up until she was standing right in front of me.

"I said, *'What are you lot up to?'*" she shouted.

I dropped my weeds and hit the Power button on the speakers. "Sorry." I reached up to adjust the cat's ears, which were sliding off my head. "We're making room for the turkeys. We'll feed all this to Pearl, then put down some wood shavings for the birds."

Amanda reached into the pen, plucking a berry and inspecting it closely. "Do you know what this is?" she asked.

"A fucking nightmare? We've been at it for hours. You wouldn't believe the size of some of these plants."

Amanda's eyebrows furrowed. "I think it's inkweed. You'd better not give it to the bloody goat. You'll kill her."

"Sugar, Honey, Ice, and Tea!" Rebecca raced out to the palm tree, where we quickly kicked the pile of death weeds away from our pregnant goat.

I let loose a frustrated sigh. "Great. Now we'll have to move all this out to the upper paddock and burn it. More work."

Rebecca drew in her breath sharply and put her hand to her mouth. Her lips started working strangely.

"What's wrong, Rebecca?" Amanda looked concerned.

"I . . . I think I just swallowed a fly."

"Ha!" I crooned. "Vegetarian no longer!"

Rebecca looked aggrieved.

"That's all right," I soothed. "I'm sure it was a happy fly."

That week, I purchased three young turkeys online. Collecting the birds, from a playful English couple who ran a bed-and-breakfast and raised turkeys for fun, was mostly uneventful. It wasn't until we got them home that the shenanigans began.

To begin with, they were surprisingly unattractive. They had dun-colored plumage that bore no resemblance to the splendid feathers I'd imagined; pale, pimply skin on their heads; and wattles that dangled like an old man's balls.

The other surprise was that turkeys can fly. I suppose I should have known this, but in fairness, I'd only ever dealt with them shrink-wrapped in plastic. I figured they could flutter a little here and there, much like our awkward chickens, so Rebecca and I had fitted a loose wire mesh over the top of the turkey pen. This proved worse than useless. No sooner had we released them into their new home than these turkeys began collapsing their bones inside their bodies, squeezing through impossibly small gaps in the wire, and heading off in all directions. One turkey crouched atop the shed, and another took off and soared effortlessly across the property, finally roosting on a shaggy old beech tree, where it gazed down at us like a smug pterodactyl. The third appeared to have an intellectual disability. He just shuffled around in the wood shavings, eating rocks and stray turds from his roommates.

"Great." I groaned, eyeballing the pterodactyl. "Now what?"

"I'll get the one on the shed," Peter volunteered. "But you guys are gonna have to figure out that other guy. He's pretty high up."

There was a stream at the back of the property, and the beech tree grew diagonally out from the bank, so our renegade turkey was doubly inaccessible: elevated about thirty feet off the ground, he was perched on a slim branch that extended over muddy water.

"I don't know what to do," I muttered to Rebecca. "Just wait here. I'll try to get to the other side of the creek."

This involved hiking out to the road, then over to our neighbor's property, jumping an electric fence, and crossing a cow paddock to make it back to Rebecca on the opposite bank. By the time I got there, she'd just about given up. But at least I'd had time to formulate a plan.

"I'm gonna chuck rocks at it till it starts flying," I called. "Then you're just gonna have to grab it."

Positioning myself at the edge of the crumbling bank, I began collecting projectiles: small stones, discarded fencing battens, twigs. Each of these I hurled at the offending turkey, trying to unsettle him enough to take off, but not to injure or kill him in the process. The turkey started making a nervous trilling sound. Over in the pen, I heard his dummy roommate trilling in reply.

"They're talking to each other!" Rebecca called. "This is good. Throw another one!"

I picked up an apple-size stone and propelled it at the tree limb, cracking the branch right beneath the turkey's perching claws. It squawked and took flight, showing off its epic strategic reasoning powers by flying right back to the pen and perching on the loose wire mesh.

Rebecca was ready. Crouched inside the enclosure, she flashed her hand out and nabbed the bird by its ankles, yanking it back inside. "Gotcha!" she hollered.

"Damn." I lay back on the bank, panting at the effort. "That is one stupid bird."

CHAPTER NINE

THE FIRST ONE IS FREE

July arrived, and with it the damp chill of midwinter. Purua's climate is wet and temperate, perfect for helping things grow. This did not apply just to fig trees and peaches. Our bedroom ceiling blossomed with mildew, wafting invisible spores on our heads. Slugs appeared in the bathroom, sliding wetly across the tiles. And a rich green slime began to coat the outside of the house, filling in the ridges on the wooden deck until the planks were as slippery as an ice rink.

Our compost heap grew moist and slimy, and one morning I found a cluster of pale, plump maggots writhing blindly on its surface. I should have known then that something fundamental in me had changed, because I did not scream or throw up, as a normal person would have done. Instead, I fetched a jar and collected the maggots for the chickens.

The growth wasn't relegated to slime, mold, and maggots. Everywhere I looked, little lambs began appearing in the pastures.

Sometimes they were so small that all I could make out from the road was a tiny pair of ears, sometimes pink, sometimes black, peeking out above the top of the lush green grass. And in Purua, lambing season meant just one thing: Calf Club Day.

Calf Club Day is a farm exhibition in which children are paraded with livestock. That sounds more depraved than it should. It's actually very sweet and educational: the kids raise their own baby goats, calves, and lambs, and on the big day, they lead them around a ring on a harness. Working farmers ask them questions about livestock care, and the winners earn ribbons with gold lettering.

But for our family, the lead-up to Calf Club Day was a nightmarish cycle of torment. This is because Rebecca and I both got hooked on the lamb.

I don't know what made me do it. Six months into our stay in the country, we already had twenty animals on the farm. At any other time in my life, that would have been crazy-person territory. Nobody sane has double-digit animals. Twenty cats in your house? Call the SPCA. Twenty hamsters, or ferrets, or dogs? You are an unemployed person who hears Jesus talking in your head.

But out here, in Purua, all those animals seemed normal. The cows, goat, and alpacas were very low maintenance, and we hardly ever saw the two cats. Two dogs didn't seem all that unusual, the chickens produced fresh eggs for our family, and we planned to slaughter the turkeys in time for Thanksgiving. The only animal who seemed at all questionable was Jabberwocky the rooster, with his scrappy attitude and insatiable libido. And after all, he was a cock. We could hardly blame him for acting like one.

But, then, that's often how addictions start. A little nip here, a little fix there, and before you know it you're on to the hard stuff. And our drug of choice was the lamb. The entire community raised lambs for Calf Club Day, which is another way of saying "But everybody's doing it." When local farmers had a sheep who died in childbirth or one who rejected her baby, they gave the little lamb away. Like crack cocaine, the first one was free.

The danger with lambs is that they're cheap and irresistible. The first time I held a newborn lamb, I was devastated by its cuteness. The big soft ears; the gentle, melting eyes; the tiny hooves; and the fluffy white tail—I didn't stand a chance. No matter that we already had twenty animals and no plan for a house in six months. I had to have one.

We put the word out in the community that we wanted to raise a lamb, and then we waited for the phone call. Rebecca spent her days running Google image searches for "cute baby lamb," and then squeaking in delight when she found a good hit. As for me, I laid in supplies. I stacked up bottles, rubber teats, and milk replacer like an anxious mother waiting to hear from the adoption agency. I cleaned out the dog kennel, piled up fresh towels for bedding, and studied books about sheep rearing. "Isn't it *wonderful*?" I babbled to Peter. "The farmers will give you a lamb for free. *For free!* Just because it doesn't have a mother." I shook my head. "Guess they're too coldhearted to rear a lamb by hand."

But no, those farmers aren't coldhearted. They just know what they're dealing with.

To begin with, there's the time commitment. A newborn lamb should be fed six times a day. That's a lot of work, especially if you're trying to do something else, such as feed a family. Shower regularly. Hold down a job.

The other thing the farmers know is that lambs grow into sheep. This may not sound like an astonishing fact, but when a lamb is hand-reared, it is unafraid of humans. Which is fine when it weighs five pounds and looks like an Easter toy. But when that lamb becomes a hundred pounds of sheep tackling you for a love cuddle, it gets a little dangerous. Of course, I knew nothing about all that, and if anyone told me, I chose to ignore them. I had a vague idea that our lamb would one day grow up, but in the early days that seemed abstract and irrelevant.

When we got the call, I was over the moon. Rebecca and I sailed out of the house and raced to Maria's sheep farm in record time. When we pulled up the driveway, there was Ba, a tiny, shivering baby taking shelter in a wooden doghouse. I lifted him up, the scent of his fleece slightly spicy like a newborn baby's smell, and he nestled into my shoulder right away. I cuddled him close, my body flooding with the warmth of new motherhood. Nuzzling my ear with his feather-soft nose, Ba bleated softly against my neck, and my boobs just about let down with milk.

That was my first taste of the lamb.

As for Rebecca, she was astonishingly nurturing. We already knew she was talented with kids, and a few weeks on the farm had demonstrated that she was equally impressive with animals. She pitched in with every job, from calf feeding to turkey wrangling, and she even made a couple of halters for Cinnamon and Lil' Lady so we could lead them around by hand.

But the reason she was so good with children and animals, it turned out, was that Rebecca didn't distinguish between the two. To Rebecca, the difference between our baby sheep and a human infant was academic. As far as she was concerned, one of them cried less and smelled like lanolin, but both should be equally treasured.

Initially, her fastidious lamb care was a blessing. A newborn lamb is legitimately fragile, so you do need to take extra care with its feeding. I begged some cow colostrum off Hamish, and Becca nursed Ba from a bottle right away. He couldn't get a latch on the lamb nipples, so I dug up the old baby bottles again, and Rebecca warmed the colostrum on the stove as she would have for an infant.

Then, on the second day, we had a little scare. Ba wasn't drinking well from rubber nipples, and he started to get dehydrated. I left to pick up Miranda from preschool, and I was just collecting her backpack when Rebecca rang me on my cell phone. "He can't poop anymore," she announced. "I think he's constipated. But don't worry. I found a solution."

"Oh, great." I covered one ear to block out the squealing cries of children. "What do we need to do?"

"Well . . . it's a little gross."

"Yes?"

Rebecca hesitated. "When they're out in the fields, the sheep mother will . . . uh, lick the lamb's bottom."

A child raced past screaming, and I leaned against the wall, not sure if I had heard her right. "Rebecca," I said sharply. "You're not actually *doing* that, are you?"

She giggled in a way that sounded not entirely convincing. "Oh, no. I was thinking maybe we could try a warm washcloth."

In retrospect, I see that the lamb-licking episode was a warning sign, but then our slide was imperceptible. Rebecca made a bed for herself on the living room floor, and the lamb was supposed to sleep in a cage beside her. But after dark, I'd hear the cage door slide open and Rebecca's soft voice as she coaxed the lamb into her bed. In the morning, I'd come outside to see the two of them cuddled beneath the blankets, tangled together like exhausted lovers.

Becca wasn't the only one who started to slip. Because Ba had to be fed so frequently, I routinely cancelled dinner parties and parent-teacher conferences. I lied to my friends, making vague excuses about housework and headaches just so I could be alone with my fix. I mean my lamb. After I fed him, I'd sink my face into his fleece, breathing in his sweet woolly smell and letting it fill me with a sense of peace and well-being.

If you've never raised a lamb, you might wonder what the attraction is. After all, it's just a baby sheep. It's got hooves and a tail. It's basically livestock. What's so great about a lamb?

To that I would counter: Ask the lady crackhead what's so great about drugs when she's living under a bridge with a glass pipe and facial sores. She doesn't see the squalor. She sees only sunshine and rainbows, peace and pleasure forever.

That's how that lamb seduced me. It was cute like a baby, but unlike my own children, it didn't whine or annoy me. On the rare occasions when I did leave the house, I found myself thinking about my lamb, wondering how cute he was being at that precise moment. Was he leaping? Prancing? Gamboling, perhaps? I felt sorry for people who didn't have lambs, because their lives seemed so gray and predictable. Unlike me, they didn't live in hyper-elevated lamb reality, full of cute things and sparkles.

Maybe that was when I started to lose it. When Ba wasn't sleeping with Rebecca, I often carried him against my body in a sling. Sometimes he walked around with a little diaper on, but even when he didn't, it wasn't really a problem. Kowhai followed Ba everywhere, and as soon as he pooped, she gobbled the turds with gusto. Even our dog was getting a taste for the lamb.

I snuggled Ba first thing in the morning, and sometimes when Peter wasn't looking, I brought him into bed with me for a cuddle. Meanwhile, our home was scattered with filth. Dirty dishes cluttered the sink; the laundry hamper vomited its contents across the floor. And I was blind to what this fluffy white toxin was doing to our family.

Yet instead of cutting down, I took it up a notch. I let the local farmers know we wanted one more lamb, and when Jackie called one evening, Rebecca and I raced out the door to collect our new baby. "She's just a few hours old," Jackie warned. "You'll have to get colostrum into her straightaway, or she might not make it through the night."

A tiny bundle of fleece pressed against Jackie's overcoat, two shiny black eyes peering out.

"D'you want me to dock her?" Jackie asked.

"Dock?" I looked confused.

"Cut her tail off," Rebecca clarified. "They take off the tail so it doesn't get dirty."

"Ah, we won't cut it," Jackie corrected. She held up a small rubber ring, about the size of a Cheerio. "We just clamp this ring on, then the tail goes necrotic and falls off. Gets a bit whiffy, but it won't harm the lamb."

I wasn't sure about that. When things go "necrotic," it's not usually a fun time. "Does it hurt?"

"Might be a bit sore. But it's quite safe." When I finally agreed, Jackie brought out a strange tool that looked like a large pair of pliers and clamped the rubber ring on our new baby's tail. She didn't even bleat, so I considered it more like a circumcision than a mutilation and promptly forgot about it.

It was dark by the time we got home. Ba bleated loudly at the sight of his new sister, and Peter rolled his eyes when he saw the swaddled shape in Rebecca's arms. He pressed his lips into a tolerant line. "I'll feed the kids," he volunteered. "You'd better go talk to Hamish."

Rebecca was already arranging herself in a nest of blankets on the couch. "It's okay," she urged. "I'll cuddle the babies." She was not referring to our children. The two lambs were sharing a blanket on her lap, and Silas and Miranda were nowhere to be seen.

It was already seven o'clock. I peered doubtfully across the road at Hamish's dark farm before picking up my milk bucket and setting off. I didn't really want to ask him for any more stuff. We'd been so picky with his milk, and he already thought we were useless city people. Tentatively, I knocked on his door.

The person who greeted me was so different from Hamish that it was as if aliens were inhabiting his skin. Showered after his long work day, he wore a clean white T-shirt and pressed gray slacks. He sipped from a cold bottle of beer as I stammered out my request.

"We have a new lamb, you see. She's just a few hours old, and we've got to get some colostrum for her, so I was wondering, I mean, I thought maybe if you have some . . ."

I trailed off lamely, waiting for him to fix me with his trademark glare. Instead, he broke into a friendly smile. For the first time, I noticed his eyes were a nice shade of hazel. "Ah yeah, that's no worries. Here, let me get a torch. I'll take you out to the sheds."

I followed his pool of light, and he led me to the back of the milking shed, where a few buckets of thick, yellowish colostrum were lined up by the wall. He picked up the first one, and even in the shifting light we both saw the bugs floating on top, the chunks where it had started to curdle. He hesitated, put it down, and went for the next pail, then filled my bucket with fresh, creamy milk.

"Hop in my ute," he suggested. "I'll give you a ride back to the road."

We bumped along the rough terrain of his farmyard, and

Hamish was actually chatty with me. "We used to raise lambs and sheep. I was a wool farmer for years. But wool's got no value anymore. People just buy the synthetics. You earn so much more off the same land now when you're farming dairy. Miss the lambs, though. It's good for the kids, to raise 'em up."

He stopped his truck, and I climbed out, stunned. "Er, thank you. Thanks!" I stammered, as he steered his truck back up his driveway.

Was it the beer? Hamish's grumpy farmer mask had tipped, and underneath he was this sweet and friendly guy.

Or maybe he was just enabling us. I named our second lamb Colette, and we called her Cou-Cou for short. She also needed to be fed six times a day, so now we were up to twelve feedings in all, with a timetable posted prominently in the kitchen. Rebecca and I took turns, but we were still measuring formula and warming milk every couple of hours, and if their feed was just a few minutes late, Ba and Cou-Cou cried out like starving toddlers.

But by now I'd developed a tolerance. I couldn't get that cuteness high from just two lambs anymore, and Becca and I started coveting a black one. Sure, the little white fluff balls were adorable, but what if we added a black lamb to the mix? And then they snuggled one another? The image this fixed in my head was almost intolerably cute. Desperate for a brand-new fix, I started calling around to local farmers.

Rumor had it that Jackie's daughter Manda had a black ram in her paddock, so I rang her up, using my most cajoling voice to win her over. It occurred to me that I sounded like the junkies on the New York City subways, the ones who "just need twenty dollars" to get home, fill their prescription, or buy a bag of drugs.

For her part, Manda was skeptical. "I've got a couple of pregnant ewes still," she conceded. "They might have black lambs, but you can never tell. Sometimes they're spotty, or they come out all white."

"Okay. Okay. That's fine." My voice sounded jittery. "But if you get any cute little black babies, will you give me a call?"

As I hung up, I heard footsteps. Autumn was on our front porch. She gave me a weird look. "Who do you think you are?" she asked. "Angelina Jolie?"

I guess I am just like Angelina Jolie. If you don't count the movie credits, or the beauty, or the humanitarian good works. We both have a crowd of ethnically diverse babies, though at last count she just had six and I was working on twenty-three. Also, she's probably got a team of highly skilled nannies, whereas Rebecca and I were stressing ourselves out just keeping up with the feedings.

Our spiral of addiction went on for weeks, and it surprises me in retrospect that I didn't recognize it for what it was. My children were neglected. My husband wasn't getting laid. My friends thought I was dead, and as for my writing, what was the point? Our lambs were all I needed, my sole source of truth and joy in the world.

Then Cou-Cou terrorized my daughter. I was slapping together dinner one night, tossing leftover pasta in a pan, when Miranda let out an ear-bending scream.

"*Mama!*" she wailed. "Cou-Cou is *dying!*" I turned to see our youngest lamb trotting happily away from her necrotic tail, which lay like a severed limb on the floor.

"Ooo!" Rebecca ran over to pick it up. "Her tail fell off!"

Sitting her on my lap, I tried to explain to Miranda that this was perfectly natural, that dead tails were as common on a farm as chicken leprosy or vampire worms. Tentatively, she reached for the tail, inspecting the dried black blood at the stump and the soft white fleece at the tip. She brushed the fleece against her cheek.

"Can I have it?" she inquired, and when we gave it to her, she tucked it carefully in her little gold handbag. I tried not to think what else she had in there.

I suppose things could have gone on like that. I can easily imagine a future in which I became increasingly obsessed with my lambs, tying bows in Cou-Cou's fleece and licking Ba's anus to cure his constipation. But then, a few weeks in, we both hit rock bottom.

One evening, Rebecca was sitting on the living room floor by the fire, crooning lullabies to her fluffy white babies, when suddenly she stiffened. "What's this?" she asked, gripping Ba's ear.

"What's what?" Peter inquired.

"Oh my God." She bit down on her lip. "I think they're ticks! What do I do?" She reached for her tablet, frantically typing in a Web search for "baby lamb ticks remove."

"Uh, put the lambs outside?" Peter spoke softly, as you would to a crazy person. "They're sheep. They get ticks. Ticks aren't a health problem in New Zealand. Having a sheep inside my house is the problem."

"Don't be ridiculous," I snapped. "We don't have a problem."

"*You're* the problem," Rebecca scowled, not looking up from her tablet. "Tweezers. I need tweezers. And a jar."

"I can get that." I hopped up from my chair to fetch her tools.

"But how will I kill them?"

"I'll put some bleach in the jar." I brought her the implements and sat back on the couch, prepared to defend our babies from Peter's boring logic. Rebecca commenced picking black specks off Ba's ears and dropping them one by one in the jarful of bleach.

"Okay, can we just consider this scene?" Peter's voice was straining to stay calm. "You're picking ticks off a sheep *in my living room*. This is no way to live. It's a health hazard."

Rebecca didn't seem to hear. She picked up the jar, peering at its contents. "You . . . *monsters*," she hissed. "How *dare you* hurt my baby?" Then she swirled the jar, giggling merrily as she watched them all die.

And that was the moment of truth. In a split second of clarity, I saw my gentle niece squatting on the floor like an animal, shrieking at a jarful of ticks.

"Okay," I conceded. "Maybe things have gone a little far. It's not that cold out tonight. Rebecca, maybe it's time for Ba and Cou-Cou to sleep outside."

She glared at me as if I'd just ordered her to eat an unhappy steak. "Fine." She clutched her babies to her chest, getting to her feet and pulling on her moccasins. "They can snuggle with me in my sleep-out. You guys are the *worst*."

But that's how it goes with addiction. Some people snap out of it; some never do. After a couple of months, Ba just didn't look cute to me. To begin with, his balls were enormous. Two huge, furry sheep testicles hung down from his hindquarters, the whole sack the size of an overripe grapefruit. His horns started coming in, and I guess these felt itchy, because he rubbed them on the trees of our property, dislodging his horn buds until his head was a bloody mess. As for Cou-Cou, she developed an unpleasantly high-pitched bleat, and whenever she saw me, she shrieked for more food.

"It sounds like she's yelling at me," I complained to Rebecca. "Like she's screaming at me in sheep language."

"Don't be silly," Rebecca swooned, her eyes glassy with lamb lust. "She saying 'I loooooove you. MaaaMaaaa, I looove you.'"

But those bleats didn't sound like love. They sounded like verbal abuse.

Then Cou-Cou started scouring, which meant I had to clean her soiled backside every morning. Despite Rebecca's research, I chose to use a washcloth. And really, that sealed the deal. Ba and Cou-cou were no longer my fluffy young babies, but instead large and unpleasant livestock.

And I knew something else. Pretty soon, we'd have to deal with the dags.

I knew about dags from Invercargill. While we were living there, I'd had the chance to chat with a sheep farmer named Spencer, and it was he who revealed the darker side of sheep care.

"Aw, sheep farming's easy, really," he told me. "Long as you drench 'em, and crutch 'em in the winter. Else they'll get daggy, and you could lose 'em to flystrike."

"I'm sorry, what?" New Zealand is supposed to be an English-speaking country, but sometimes I wondered.

"Crutch 'em," Spencer repeated.

"And that's . . . giving them medicine?"

"Nah." He looked impatient. "That's drenching. Crutching is shearing round the backside and clearing off any dags."

I couldn't quite process what he was saying. "You're telling me you have to shave a sheep's ass?"

Spencer smiled. "Got to. It's a health thing. Else the maggots get in, and they'll eat the body of the sheep. Got to take all that wool off. It's a bit uncomfortable round the dags, because the maggots can be quite large."

I swallowed. "And what are . . . dags?"

"The stray bits of shit that get stuck in the fleece."

That was a lot of reality. And back then, I didn't even own a sheep. But now I was facing an imminent need to crutch and de-dag. After all our farming traumas, I thought I could handle anything. But I wasn't ready for ass maggots. Or the regular ovine Brazilian.

That night I climbed back into bed with Peter. "Is it finished?" he asked, snuggling into me. "No more lamb licking?"

The thought made me physically ill. "No more lamb licking."

"And we can kill them for food?"

"Absolutely." The spell was broken. "No more glass pipes under a bridge for me."

Peter drew back. "What?"

"Nothing," I reassured him, pulling his arm around my waist. "Go to sleep."

The next morning, I was covered in pee. This was not a withdrawal symptom. Sometime in the night, Miranda had slipped into our bed, fallen into a deep sleep, and peed all over us.

The trick when this happens is to lie perfectly still. As long as you don't move, the pee remains at body temperature and you can almost pretend it never happened. The moment you stir, you've blown it. The pee cools down, the illusion breaks, and you realize you've slept in a toilet.

I rolled over and winced. "*Miranda!*" I hissed. "Did you have an accident?"

"I'm sorry, Mama," she whimpered, beginning to cry. "It wasn't on *purpose*. I want to be your *friend*!"

Which made me feel like a cruel mother, and then Peter and I had to get up, strip the bed, change our clothes, and clean up the pee, all the while saying sorry to *her*. Which is a great metaphor for parenting, I think: you do the work, they pee all over you, and then you have to apologize.

Then Silas wandered into the room. "Good morning, son!" Peter opened his arms wide for a cuddle. "Did you sleep well?"

"Bus," Silas replied, with utmost seriousness. His eyes are so dark and thoughtful, and he gets a look of such profound intelligence on his face, that sometimes I'd swear it was me who was limited and Silas who was way ahead of the rest of us.

"Would you like some breakfast?" I ventured, but he turned and strode out of the room.

"Bus," he reiterated from the hallway.

"That kid's an alien." I sighed and slouched back on the pillows.

"Yes, but he's *our* alien," Peter reminded me. Then we gathered the pee-soaked linens and went out to check on the cows.

I didn't spend much time mourning my love affair with the lamb, because in the next few weeks Pearl took up most of our focus. By now the pregnancy was obvious. Pearl was most definitely with kid.

We knew our goat was either pregnant or psychotic because now she'd do anything to get at the good food. First she managed to slip her chain and circummasticate the lemon tree, denuding it of bark and dooming it to an untimely demise. She developed a taste for hydrangeas and began rearing up on her hind legs to reach the tallest blossoms, clattering her hooves on the windowsill by my office and freaking me out while I was trying to write. And one night, when we'd shut her up in the chicken pen, making a soft bed of hay so she could curl up out of the rain, she smashed her way through a plywood wall to get to the grass outside.

Now, I've been pregnant. Twice. I've had my share of cravings. I once ate an entire jar of cornichon pickles, then drank all the brine, because I was just that crazy with baby hormones. But you didn't see me breaking down walls with my face. That's a step beyond. That's goat crazy.

And goat crazy seemed to work for Pearl. After she splintered a hole in her pen, she ate her fill until returning, plump and self-satisfied, to our front porch for the night. She grew sleek and glossy, swollen with the promise of new life growing within. I found myself forgetting all about my fever dream of lamb lust and I encouraged Rebecca to spend more time with Pearl, to stroke her and make her little salads from the garden.

All that changed when I started reading about goat birth, or "kidding," which is what they must have been doing when they described what to do if a baby goat gets stuck in its mother. Margaret Hathaway, in her book *Living with Goats*, isn't afraid to get explicit: "If the goat appears to be having difficulty expelling the kid," she suggests, "it is a good idea to help the kid out with a gloved, well-lubricated hand." She continues: "Gently inserting two fingers into the doe's vagina..."

"Oh, *hell* no," I told Autumn and Patrice. "I'm not going anywhere near a goat vagina, glove or no glove."

Our friends exchanged a look. Autumn grew up in rural New Zealand, and while Patrice was learning how to sear foie gras in fine restaurants, Autumn was helping her parents around the dairy farm. Patrice shot me a meaningful look. "I have seen Autumn with her arm up to *here*"—he pointed to a place halfway up his bicep—"inside a cow's vagina."

"No gloves, either." Autumn giggled, clearly enjoying my horror. "Just slipped right in."

Chastened, I looked down at Hathaway's goat book, which listed a number of "kidding kit essentials."

> *Lubricant.* I now knew what that was for, though I wished I didn't.
> *Shoulder-length gloves.* Shudder.
> *Obstetric leg snare.* Please God, don't tell me.
> *Shot glass.*

"Ha!" I pointed to that last one in triumph. "Something's finally making sense!"

Autumn raised an eyebrow. "I think that's for iodine," she corrected me. "You know, so you can clean the umbilical area after the birth."

The way things were looking, I'd be happy to do a shot of iodine. Right now. But luckily, the kidding was still four months away.

Meanwhile, the injured chicken seemed well enough to join the flock. Her wings were fairly even, and she was moving around without too much trouble. One day, I let her out of her cage to stroll around the garden.

"I think she's getting better," I observed, watching her peck at the grass. "Look at that. I think the Chicken Hospital really worked."

Peter was sitting with Silas, who was dipping a metal goblet in a mud puddle. "You think she's better?"

A feeling of satisfaction came over me. "Yeah. I love that our animal had a problem and we figured out how to fix it." Silas, who had evidently decided the mud puddle was full of chocolate milk, sampled the water in his cup. He grimaced and spat it out.

"That's great, baby." Peter approached the chicken, crouching down to her level. She waddled away, but he scooped her up.

"What are you doing?"

"I want to see my chicken fly." Then, before I could stop him, he tossed the chicken in the air. At that moment, several things happened in quick succession. The chicken fluttered her good wing like crazy, but her other wing hung limp, so she spun like a football and dropped like a stone.

Then Jabberwocky pounced on top of her, fluttered six times, and walked away to eat worms.

"What just happened?" I asked in horror.

Peter looked pale. "I think he just fucked her."

Miranda came up behind me. "Is *fuck* a grown-up word?"

"Why'd you throw the chicken?" I asked accusingly.

"It's a bird. You said it was better."

I shook my head in disbelief. *Who throws a sick chicken?* "Well, at least there's one thing we can be thankful for."

"What's that?"

"It wasn't a duck."

CHAPTER TEN

HEAVY BREATHING

We weren't the only ones with lamb trouble. All the families in Purua adopted baby animals for Calf Club Day, and challenges inevitably arose. Autumn's daughter Nova chose a pale gray lamb named Cardigan, and when they went out to feed her one morning, they found her dead. No one knew why. She was just too weak to make it through the night.

"Those girls must feel so awful!" Rebecca crooned when she heard the news. "Let's invite them to breakfast, so we can try to cheer them up."

Like most displaced New Yorkers, Rebecca had spent much of her time with us bemoaning the lack of bagels in New Zealand. Never mind the world-class scenery and the pageantry of exotic birds in these islands. What Rebecca needed most was a sesame bagel with a schmear. We decided to try baking our own, and on Sunday we invited Autumn, Patrice, and their children over for a taste.

"What's that?" Maris asked when the first tray of bagels came out of the oven. We'd made sesame, garlic, and everything bagels. Plump and steaming on the cookie sheet, they looked like a taste of home.

Rebecca went pale. "You've never seen a *bagel* before? They're only the best thing ever!"

"Why does it have a hole in it?" Nova wanted to know.

"Stop asking so many questions," Autumn ordered. "Why don't we show Antonia what you brought?" She reached for her bag and pulled out a tablet.

"Ooo, pictures?" We gathered round the tablet, anticipating photos of smiling children and lambs.

"We made a movie," Maris mumbled, looking up at me with big, dark eyes. Nova just smiled.

Autumn hit the Play button.

On the screen, Maris was bustling about the kitchen pretending to make supper for her family. She opened a recipe book, stirred something in a pot, tasted and seasoned her soup. Carelessly, she wiped her hands on her apron and turned back to her recipe.

And then an ominous soundtrack began. Behind Maris's back, the pantry door began to move. A slender white hand reached out, sliding the pantry open and reaching for Maris's neck.

In the following scene, Maris lay broken on the floor. Her throat was slashed, blood oozing from the side of her mouth. The camera zoomed in on her neck, showing flaps of skin peeling back from the open wound.

In the final scene, the body was gone. Autumn came home, hanging up her coat and moving to the kitchen, not noticing the blood that dripped from the walls. The pot was still simmering on the stove. Autumn took a spoon and tasted. "What a marvelous stew!" she exclaimed. "It's so rich and meaty!" Then: "I wonder where Maris is?"

The movie ended. Peter gave an uncomfortable cough-laugh.

"Oh, *wow*," Rebecca stammered, pushing her bagel away.

"They're really enjoying that theater blood you gave them," Autumn explained.

"And you're doing such a great job!" I cheered the girls. Since no one had much of an appetite for bagels anymore, I decided to change the subject. "I hear Cardigan didn't do so well," I told Nova sympathetically.

She smiled. "It's okay. We got a new one."

"Great! What's his name?"

"Pixie."

"Aw, that's so cute!" Rebecca clasped her hands. "Do you have any pictures?"

"No. He's dead, too."

"We are thinking we will try a calf next time," Patrice suggested. "The lambs, they are not going so well."

Nova continued to smile, and for a serial lamb killer, this looked a little evil.

"You should try the cream cheese with chives," Rebecca offered, to fill the silence. "That one's my favorite."

None of our animals had died of natural causes just yet, but it was time to start slaughtering some. The turkeys had proved to be disappointing pets. We were afraid to let them free-range, lest Kowhai help herself to a pre-Thanksgiving snack, so they stayed in their pen, eating expensive food and producing copious amounts of manure.

The birds themselves had cost just twenty dollars each, but we were feeding them something called Meat Bird Crumbles, a high-protein feed that cost the same as dehydrated steak. "Those birds had better pace themselves," I warned. "They might not make it to Thanksgiving."

But it was still only June. Toward the end of the month, I started e-mailing Michiko's lawyer about buying her property. It turned out she was eager to sell, just as Amanda had predicted. I didn't understand all the details, but it seemed a bankruptcy or a foreclosure wasn't the best calling card for a professional accountant. And yet, that husband made me nervous. The police didn't have enough evidence to charge him yet, so he was still living nearby.

"He has to sell," Michiko wrote through her lawyer. "He has no ability to pay. And I think there is no harm to your family."

That was pretty tepid reassurance, but we felt we were running out of options, so we decided to move ahead with the purchase. We agreed on a price and submitted our offer. Everyone in the community seemed pleased with our decision, since we'd get to stay in Purua and help out their friend at the same time. Autumn even called to congratulate us. "You'll soon be the biggest gorse farmers in Purua," she teased. New Zealanders have a special kind of hatred for gorse, so this was sort of like congratulating someone on his new swamp full of poison ivy and alligators.

Then one day I got a call from Kim, the real estate agent who had shown us Martin and Klara's property. "I really shouldn't be telling you this," she whispered over the phone, "but Martin's had a surgery on his eye. He's going about with a patch now."

I thought of that sixty-acre section, all steep hills and native bush. "He has one eye? How does he get around?"

"I really don't know," Kim replied. "But I think they'd like to meet with you. They're ready to come down in price."

"You want to live in that crazy house?" Peter asked when I told him about the Slovakians. "You want to power your washing machine with a windmill?"

"I don't know." I shrugged. "Maybe. That land is so gorgeous."

There were swimming holes, open meadows, secluded glades of trees. So what if the house was illegal and the whole thing was powered with a windmill and happy thoughts? We could work with that.

So Peter called Kim, and she notified Martin and Klara that we wanted to meet. I have no idea how she got in touch with them, since they had no phone, Internet, or street address. It's possible she employed a carrier pigeon.

That afternoon, I took the kids to visit Amanda. Our visits to Amanda's house were always relaxing, because the children all ran in a pack, which meant Amanda and I could sit in the kitchen drinking wine and scarfing down potato chips. Silas loved it, too. There was a trampoline he liked to jump on, and he'd mostly stopped pooping on their lawn. He'd even used the toilet once or twice.

The only problem was that the conversation always revolved around Pearl. Now that we knew she was pregnant, everyone wanted to talk goat babies. Especially Amanda. She considered Pearl's babies to be her grandchildren, since she'd reared their mother by hand, and she never seemed to tire of the topic.

When I sat down at her kitchen table, Amanda started right in. "Have you been fondling her teats?" she asked with a sly grin.

I had a chip halfway to my mouth, and I dropped it. "Have I *what*?"

"Fondled her teats. Oh, you'll have to fondle them. Helps with the milking later, so she gets used to you. You'll have to get in there and have a good fondle."

"Could you not say *fondle*?"

"Why don't you want me to say *fondle*? What's wrong with fondling? I like to get in there and have a good fondle."

"*Please*, Amanda."

She grinned. "Okay, no fondle. What should I say instead of . . . *fondle*?"

I felt a little sick. "Manipulate? No. Manhandle. I think I could cope with *manhandling* her teats."

Soon after, I collected my kids, making the excuse that I had to go home and get dinner in the oven. But Amanda had me thinking—all this talk about getting fresh milk and cheese, and I still hadn't touched an udder. What if I couldn't do it? What if Pearl kicked me in the face, or I had some kind of udder allergy that made my skin break out in hives? What if it just grossed me out to squeeze a goat's boob?

So the next morning, I went out and tried to manhandle Pearl's tits. I mean teats. Touching a ruminant's teats is not something I ever had occasion to do in my city life, and they weren't quite what I'd expected. To begin with, they were surprisingly large. I thought goat teats would be about the same size as a rubber baby nipple, but that was hopelessly wrong. Pearl's teats were like two large, uncircumcised penises dangling close to the ground. One of them was much larger than the other, a sort of giant, swollen goat breast. Plus they were speckled. Pearl's teats were a pale shade of pink, with gray splotches up and down their length. The whole package was intimidating.

Also, I had no idea how to approach her. Should I coax her to me or try a sneak attack? I went with the element of surprise.

"Hi, Pearl," I said, in a tone that was intended to be soothing but just came out creepy. "Can I get in there and have a fondle?"

She didn't scream or run, so I touched the big, speckled teat with one hand. Instantly, she jerked away. Which I guess is what anyone would do if a stranger came up behind her and grabbed her by the tit. And that's what surprised me: that a teat really does feel a lot like a breast. It was warm and soft and surprisingly sensual. I liked touching it a little more than I should have. Pearl must have sensed this, because she trotted away to the safety of her favorite tree, then stood there in the shade looking miffed.

"Mama!" Miranda came running, this time completely nude apart from a stuffed blue monster tail she'd tied to her waist with a ribbon. The tail was about six feet long and filthy with mud and clumps of grass. "Kowhai did kill a chicken!"

"Come on, Magnolia," I chided, trying to act like I hadn't just been feeling up a goat. "Remember last time, the chicken was just hurt, and then we made it better."

"No, no, really! Come with me! I'll show you!" She dragged me down to the chicken coop, and there was Kowhai, in her favorite spot beneath the flax bush. Feathers were strewn everywhere, and she was definitely chewing on a corpse. It looked extremely chickenish.

"God*damnit*!" I swore, chasing Kowhai away. I glanced at the henhouse door, noting quickly that the latch was broken. Kowhai must have just strolled inside for an early snack. Miranda and I crept forward. This chicken was most definitely dead. There were crucial parts missing, such as the face. Blood was spattered everywhere.

"Rebecca!" I called.

Becca came over, took one look at the dismembered chicken, and went pale. "I can't deal," she announced. "I don't do well with dead things. I need to go lie down." She wandered back to her sleep-out, where she probably lay down on her acupressure mat, which was covered in hard plastic knobs. She claimed they relaxed her. I liked to call it her nail bed.

"What do we do with the chicken, Mama?" Miranda wanted to know. "Do we eat it?"

"I don't know," I shrugged. "Let's get a bucket."

But a bucket was only part of the cleanup job. The next step was to call Skin. "He'll know what to do," I told Miranda, feigning confidence.

"Aw, that's easy." Skin laughed when I got him on the phone. "Ya tie it round her neck."

"I . . . sorry? I tie a dead chicken on my dog?" I thought about this for a minute. "But what do I tie it with?"

"A knot."

"Yeah, but—won't the chicken, I mean, won't it rot and fall apart?"

"Ya can put it in a bag if ya want. But then you leave it on there for a week, and she'll never do it again. Works a treat. Might want to keep her away from the house, though. Gets a bit whiffy."

When Peter got back from work that evening, he was startled to find his dog crouching under a lemon tree, a chicken corpse lashed to her neck with twine. I'd stuffed the hen in an orange mesh bag first, so that pieces wouldn't fall off as it began to decompose.

Peter emerged from the bedroom, where he'd changed into his farm clothes. He peeked out the window, inspecting his dog. "You're sure about this?" he asked tentatively. "It seems kind of cruel."

Kowhai certainly looked miserable, a canine Ancient Mariner in disgrace. She cowered beneath the tree, too uncomfortable to sit. Her tail was pointed straight down, an anchor line sinking between her legs.

"Not as cruel as getting shot by a farmer. She can't chase chickens, or sheep, or cows even. Skin says if she does, she'll get a bullet in the head."

Peter looked doubtful. "We'll try it one night. I might take it off in the morning."

The next morning, when I woke up, Kowhai wasn't barking, and

I went outside right away to check on her. My first thought was that she might have strangled herself on the twine I'd tied around her neck, but she was safe, and actually quite pleased with herself. Overnight, she'd gotten into the orange mesh bag and devoured the rest of the chicken. All that remained were two chewed-up feet. Kowhai stared up at me, her face full of guilt and repentance. "It's all right," I comforted, releasing her from the lemon tree. "It's not your fault."

I told Peter what had happened. "What did you expect?" he asked. "You basically rewarded her with a sackful of food."

I sighed, looking over at Silas, who was sitting on the couch flipping through the pages of my *New Yorker*. "Bus," he announced, pointing to a cartoon. "Bus."

I walked over and sat down next to him. "No," I said gently. "That's a man. Can you find his eyes?"

"Bus," Silas replied.

"That's all he says anymore," Peter pointed out. "Have you noticed? Bus. Bus-ah-bus-ah-bus."

Silas put his hand over mine. Very seriously, he looked up into my eyes. "Bus," he repeated.

I had no idea what he was saying.

"Mama?" Miranda wandered in, naked with her shiny black gumboots and a gold purse over one shoulder. "Can I please have a chunk of cucumber?"

"Sure." I got up off the couch, wiping my eyes with the back of one hand. Sometimes it floods me: I picture Silas as he'd be if he weren't disabled. He'd be building forts, mixing zombie poison, asking questions about the chicken tied around his dog's neck. I shut my eyes and breathed, composing myself before turning around.

"Don't you go meet with the Slovaks today?" I asked Peter,

pulling a cucumber from the refrigerator. "Why don't you see if you can go buy us a house?"

Peter laughed, which is what he does when I'm being completely unreasonable. "What about Michiko's place? We've got an offer in on that one."

"Yeah, well, she hasn't accepted it. If we get a better deal with the Slovaks, we can pull out at any time."

It felt good to have extra options all of a sudden. Peter went off to get dressed, and I sat down with my e-mail.

"Bus," Silas demanded. "Bus-ah-bus-ah-bus."

I looked at him, concerned. "Bus? There's no bus today. It's Saturday. Wouldn't you like to have some breakfast?"

"Silas doesn't want breakfast," Miranda interpreted through a mouthful of cucumber. "He just wants a bus, Mama."

I got up to make him a peanut butter sandwich. "Well, he might not be able to talk so much, but he still has to eat."

Peter emerged wearing a worn orange T-shirt and jeans. "You want a house, baby? I'll go buy you a house." Peter loves making deals. With a happy grin, he kissed me on the cheek and took off.

"So," Miranda said when Peter had left, "what do we do now, Mama?"

"I don't know what you're doing, but I am sitting. And checking my e-mail."

This might have been a mistake. Gay and Mike, our alpaca breeders, had sent me a series of messages, each with the subject line "Orgling." I clicked on the first one. It contained several explicit YouTube clips of alpacas in intimate embraces. Here were the teddy bear camels thrusting away at one another, a kinky Peruvian subculture that now I couldn't unsee. "I wanted you to know that alpacas don't just spit!" read Gay's chatty e-mail. "They also orgle, which is the sound they make when they mate." Then she closed: "Alpaca porn, whatever next?"

From the kitchen, I heard heavy breathing. "What are the 'pacas doing?" Miranda wanted to know. "In that movie?"

"Orgling," I muttered, closing the e-mail.

"Oh." Miranda nodded sagely. "Orgling. That's good."

I went into the kitchen to see what was happening. Silas had three different pot lids down on the floor and was spinning each one separately. When one slowed down, he gave it another spin. And he was breathing heavily, almost hyperventilating.

"Silas?" I asked, bending down. "Are you okay? What's up with the pots?"

"Bus," he replied predictably. Then: "Pah-mmmm."

"Pot? Are you playing with the pots?"

"Pah-mmmm. Bus." He gazed up at me then, but his big brown eyes didn't look urgent and sharp anymore. They seemed far away.

My laptop chimed. I went to check my in-box. "What is it, Mama?" Miranda asked. She was trying to stuff the rest of her cucumber into her little gold evening purse. "More 'paca movies?"

"No." The message was from Michiko, through her lawyer. I clicked on it.

> Duncan has said he will not sell for this price because he wants more money. This is a very shame result. So sorry.
> Michiko.

"*Shit*," I swore under my breath.

"Mama," Miranda called from the kitchen. "I think Silas is orgling."

After no more than an hour, Peter's car pulled into the driveway.

"Papa's home!" Miranda rushed outside to meet him. I knew it couldn't be good that he was back so soon. Through the window, I saw her offering him the mushy stub of cucumber from her bag. Peter smiled and pretended to eat it.

"So, Master Dealmaker," I greeted him when he walked back inside, "how did it go in Slovakia?"

Peter sat down at the dining table, gratefully accepting a glass of water. "Strange," he said. "Very strange. The guy's on a respirator, he has a patch on his eye. I thought he was gonna keel over right there while I was talking to him."

"Yeah?"

"They had it all worked out. They took twenty percent off the price. Well"—he smiled modestly—"fifteen. I dropped them down to twenty."

"That's fantastic! When can we move in?" I gave him a big kiss on the mouth.

"There's just one thing." I waited. "They want us to cut Kim out of the deal."

"*What?* Are they high? We can't cut out the agent! It's illegal!"

"What's 'high'?" Miranda asked.

Peter shrugged. "He said if we wanted to worry about agent fees, we could just—"

"—go live in the suburbs," I finished for him. "Well, that's the end of *that* house." I flopped down next to him, pressing both hands to my forehead. "God, is anybody *normal* out here? Fondling goat tits, tying dead chickens to dogs. Michiko's deal fell through because her creepy husband wants more money! And now the one-eyed man—"

"He still has his eye I think," Peter corrected. "It's just got a patch on it."

"*Patch*-eyed man who wants to rip off the agent. What are these people *thinking*? Is everyone out here completely insane?"

It was a good question. Miranda had wandered outside, still with no clothes on, probably to chase a chicken. Silas was still hyperventilating in the kitchen. And we were so busy complaining that no one remembered to check on him.

The southern winter passed without further mishap. We tended our animals and kept searching for homes, sipping fruit wine by the fire at night. But at the start of October, I finally put my foot down. Adding up the cost of the feed we'd lavished on our three turkeys, each bird now cost about eighty-five dollars, and they were still strolling around in their pen, trilling and eating our food.

Peter called up Skin. "If you could just show us how to butcher them," I heard him asking. "If you could just show us the first time, Antonia and I really want to learn how to do it ourselves. Do you think you might have some time? Like on Saturday?"

Apparently he did, because that weekend Silas was riding his trike around the palm tree when Skin's gold sedan rolled down our driveway. Silas stared as Skin got out of the car and pulled off his oilskin, balling it up and shoving it in the passenger-side window.

"Got an axe?" he asked.

This wasn't usually what people said when they came round for a visit. "Uh, yeah," Peter said, heading out to the garage for the tool.

Skin turned to me. "Normally you don't kill them in any month with an *r* in it. That's when they eat the crickets, and the meat goes off." He started walking toward the turkey pen, nodding at the birds. "But you've got 'em in the pen here, so she'll be right."

"No crickets, I don't think," I stammered. "Just really expensive feed. I'm getting sick of paying for it, actually."

Peter returned with the axe, and Skin pointed to the back of the house. "We'll do it over this way. Derek's got some old tree stumps round the side there. Make a good chopping block."

"Mama, what are you doing?" Miranda wanted to know, emerging from the deck with That Baby wrapped in a sling. Ever since she'd seen me carry Ba around the house, she'd wanted to wrap her dolls in baby carriers. I hoped she wasn't treating them for constipation.

"Well, Magnolia, Skin's gonna show us how to cut a turkey's head off," I explained, hoping she wouldn't be too upset.

But Miranda was a country girl, and she immediately saw the possibilities. "And if Jabberwocky is too mean to us, then we will cut his head off, too."

"Jabber what?" Skin asked.

"Our rooster," I explained, rolling my eyes. "She wants to cut its head off."

"Yes," Miranda told him solemnly. "And we will never put it back on."

"Good on ya, Miranda," Skin laughed. "You show that rooster who's boss." He stretched the turkey's head out on a tree stump and raised the axe in one hand. With a crushing blow, he hit the turkey square on its neck, sending the head shooting across the grass.

"Go fetch that head, will ya, love?" Skin gestured to Miranda. When she scampered off, he muttered, "Can't stand roosters meself. Angry bastards. Fight first and talk later. 'S no good."

Miranda brought the turkey head back just as Skin was setting aside the carcass. Her hands and wrists were slick with blood, but it didn't seem to bother her. "See this?" Skin asked, taking a head in two hands. He pulled a flap of muscle at the base of the skull, making the turkey's beak open and shut. In a rasping falsetto, he squeaked, "Hullo, Miranda. You're a very lovely girl. Would you like to play with me?"

"Oh, *gross*," Peter said, groaning, but Miranda's eyes grew wide with joy.

"Wow!" she crooned. "Skin made that turkey talk!"

Skin picked up the next bird, chopping its head off as easily as the first. "Fight first, talk later," he continued, as though there'd been no interruption. "I used to be like that. Can't go to town no more."

"Why not?" Peter asked.

"Aw." Skin shrugged, then chopped the head off the third turkey. "Every time I go to town I get arrested."

He sat on a stump then, handing each of us a warm turkey carcass. "Here," he offered. "Let me show you how to pluck 'em. You'll need to know, once you be killing that rooster of yours."

Skin stuck his hand up a hot turkey's ass then, just as easily as if he were slipping on a glove. Wincing, I followed his lead. I thought it would be disgusting. I wrinkled my nose, bracing for a wave of nausea. But it felt surprisingly delicious, like the inside of a warm cherry pie. Except with no cherries and more internal organs. We sat together on stumps, pulling guts from our turkeys and chatting amiably about life. The evening sun lit everything in a pumpkin-hued light. Jabberwocky crowed in the distance, the leaves of the great totara tree rustled, and it was surprisingly lovely to sit there and eviscerate birds.

Skin talked about his daughter Amber, who was coaching sports for special-needs kids. Shyly, I told him about Silas.

"Ah, kids'll be who they are, not a thing you can do about it," Skin observed. "Take our Liam. That's our nephew on Lish's side. Born delayed, just like your Silas. Doctors said he'd never walk. But now he runs just like the rest of 'em. Doesn't talk like they do, but that's just him. 'That's our Liam,' we say, an' we love 'im."

Silas and Miranda crouched near the guts pile, poking at sludgy things with sticks. Our two cats paced back and forth on the thick wooden fence, hoping for scraps.

"This here's the gizzard," Skin showed us, slitting open a mysterious crimson organ with his knife. I was startled to see rocks inside, along with a handful of corn.

"That's how they get the nourishment, see?" Skin pointed inside with the tip of his blade. "Turkeys eat the stones, and the hard rocks grind up the maize. Makes it digestible. Anyways," he flicked the gizzard over his shoulder. "Cats love 'em."

It occurred to me then that in a lifetime of cooking and eating animal meat, no one had ever shown me how to slaughter and prepare my own food. It was a more useful skill than French verb conjugations or the canon of Western literature I'd studied in college. When I first met this guy, I'd thought he looked like a pirate. Now I felt lucky to know him.

Before he left, Skin hung up the turkeys on the back deck, explaining that they'd have to bleed overnight before we could freeze them. Peter walked him out to his car.

"D'you want to know my real name?" Skin asked Peter, right before he left.

Peter feigned surprise. "It's not Skin?"

"Nah." Skin balled up his jacket and stuffed it through the open passenger-side window. He got in behind the wheel. "That's my nickname. I was a twin, and I was always the skinny one, so they called me Skin. Real name's Dennis." And with that, he took off.

Peter reported this bit of knowledge back to me in the kitchen. "Doesn't seem like a Dennis," I commented. "Maybe that's why he changed his name."

"Yeah." Peter laughed. "That and the part about getting arrested. Both good reasons to change your identity."

I guess Skin enjoyed Peter's company, because the next day he called us at eight in the morning to invite Peter out to play.

"Be doin' some pigs out Pipiwai way," Skin offered by way of explanation.

Peter wasn't sure about this. "He must mean slaughtering them, or maybe butchering them?" he wondered as he hung up the phone. "And I like bacon. I eat the stuff, so I should learn what goes into it, right?"

So he did learn. And what goes into pig farming, apparently, is flying nut sacks.

The day was devoted to pig castration, an athletic endeavor in which two men hold up a series of screaming piglets while some guy slits open their scrotums.

Peter made the next part sound highly scientific. "He lops off the nut and flings it," he told me later, looking pale and slightly traumatized. "There were nut sacks flying everywhere. One hit me in the head, one bounced off my arm, and I had nuts sliding down my gumboots." He reached for a beer. "We can never get pigs."

"Really?" I whined. "But what about making our own bacon? And sausage?"

"They are disgusting, loud, and horrible. I'm glad I went out there. I wanted to come to terms with my own hypocrisy around food."

He took a swig of his beer. "And now I'm fine with it. We are buying bacon at the store."

CHAPTER ELEVEN

SPINNING AND SPITTING

Animals need simple things. You feed them, you water them, you give them fresh grass and shade from the sun. Each day, when I woke up, I did my rounds on the farm: putting out biscuits for the dogs and cats, bringing kitchen scraps and pellets to the chickens. Rebecca took care of the big animals, checking in with the alpacas and cows. Ba and Cou-cou had been weaned by now, but she still visited them each day, keeping them company and singing them songs. I was even starting to enjoy Jabberwocky. We kept the chicken feed in a garbage bin, and when I went down to feed him, I brandished the lid like a shield.

"You want a piece of me?" I'd growl, wielding my red plastic armor. "Try it, rooster. I'll pull your guts right out your ass." I meant it, too. Now that we were starting to learn some skills on the farm, I wasn't so scared of that eighteen-inch punk.

Weeks passed, and soon we were all on baby watch. Pearl's belly was heavy now, her sides plumped out and her attitude atrocious.

Far from letting anyone have a fondle, she wouldn't let us even approach. Peter tried to pet her, and she reared up, her ears on high alert, as though he were a dog and he'd tried to bite her.

The problem with this was that we actually needed to examine her, to figure out when she'd give birth. The most reliable method is to feel for the ligaments at the base of the spine; when these soften to the point where it seems they've disappeared, her kid will be born in a few hours.

But no one was touching those ligaments. Get within five feet of Pearl, and she'd edge away, her udders and giant penis-teats swaying painfully between her legs.

"She looks so uncomfortable," I told Peter. "It's got to be soon."

"Did you get all the stuff for the birth?" he asked.

"Of course," I assured him. "We have everything."

This was a blatant lie. The truth was that the only obstetrical tools I had on hand were a half-used jar of Vaseline and a bottle of iodine. The Vaseline was for greasing my hands if I had to, and I still wasn't exactly sure what to do with the shot glass of iodine. I'd looked at some specialized birthing gear in the farm store, but there was no way I would know how to use it. Pearl would enjoy a natural birth. With a couple of amateurs in charge and not a veterinarian in sight, she didn't have another choice.

Since she wouldn't let us touch her, we had only one other way of estimating the delivery date. We had to examine her hindquarters for the clear, viscous discharge that signals the start of labor. This became something of a game in our household.

"Well, I'm off to look at Pearl's vagina," Rebecca sang out one evening while I sautéed some onions and garlic for supper.

"Any mucus?" I asked when she returned.

"Not yet," she reported. Then, amazingly, we all sat down to eat.

Of course, the reason the animals' needs consumed us was that they were all so easy. I could feed a hungry dog. I could fight an irrational rooster. I could even see myself pulling babies out of a goat's vagina, if it came to that. What I couldn't understand was my son.

Over the past few weeks, Silas's world had constricted. His vocabulary had shrunk down to *iPad* and *bus*, and he chanted these words like an incantation. He lost interest in building towers or playing with cars and now spent his time arranging strange items: the empty tin cans from recycling or stray bits of dried macaroni. He placed these in mysterious patterns, making cryptic crop circles that no one understood. And if anything went wrong—a macaroni out of place or a cabinet door he couldn't pry open—he'd pant, taking deep, labored breaths that made him sound as if he were gasping for air.

I asked Nick about it, because he'd worked with Silas as a therapist. Nick took a few moments to watch him one night after supper, while Silas arranged plastic sippy cups on the kitchen counter. They weren't arranged by size or color, but he was careful to turn each one so the cartoon character was facing the same way. As he reached to adjust one, it tipped over and fell, sending Silas into a full-blown panting attack. He bent to pick up the cup and replaced it, breathing heavily the entire time.

"I've never seen anything like it," Nick said finally. "I'd guess it's just mood regulation. When you and I are frustrated we just take a deep breath and carry on. But maybe Silas needs more sighs to get the same soothed feeling."

He paused. "I'm guessing, really. Actually, I don't know."

"I don't know" was the opinion experts usually gave on Silas, whether they were doctors, geneticists, or shrinks, so eventually we just stopped asking. Instead, we distracted ourselves with other things. The flood of alpaca porn in my e-mail, it turned out, was a lead-up to the annual alpaca shearing, and I'd been looking forward to this day for months. Alpaca fleece is the payback for all the creatures' attitude and slime. It's extravagant and sought-after because it's so soft: touch a handful of fleece and it seems you're holding nothing at all.

So I threw myself into shearing preparations. I still wanted to knit an elegant pashmina with McTavish's chocolate-brown fleece, and to do that, I'd need some new skills.

"Will you show me how to, uh, spin?" I finally asked Rebecca one morning.

"Of course! I just thought you weren't interested." That was a reasonable assumption, since I hadn't dared touch a handful of fleece since the Day of the Indian Chili Bomb. But I was hopeful the spinning would go better than the picking. At any rate, it could hardly go worse.

Before I met Rebecca, I thought spinning wheels existed only in Disney cartoons. But Rebecca's wheel was an actual tool made from wood and metal parts. Though it looked like it could turn straw into gold, she used it to produce yarn.

Rebecca dragged her wheel into the living room and started the lesson. First, she pointed to a wooden stick-thingy. "This is the Mother of All," she explained.

"The Mother of Who?"

"The Mother of All." She smiled sentimentally. "I love those old-fashioned names, don't you? Now, the Mother of All mounts the Maidens—"

"You're kidding, right?"

"—and they lead the yarn to the Orifice—"

"I thought this was wholesome!"

Rebecca flushed. "I don't know. I never really thought of it like a sex thing. They're just the traditional names for all the parts. Should I continue?"

"Well, I don't know. I'm a little offended. I feel like I should wash out my ears or something. Where's the prick?"

"The what?"

"You know, the prick. The one Sleeping Beauty pricked herself with and passed out for like a hundred years?"

Rebecca shook her head. "There is no prick. That's totally made up."

"So you're telling me there's all these Mothers mounting Maidens and their Orifices with no prick?"

"Um. There's a knob. Does that help?"

"Whatever," I sniffed. "I don't know about this spinning thing. Maybe I'll just concentrate on shearing."

I'd been to a sheep shearing once before, so I was looking forward to seeing how they did it with alpacas. Unlike that filthy spinning business, sheep shearing is athletic and beautiful—even poetic, if you're in the right mood. When a skilled shearer grabs hold, the beast goes limp in his arms, as though it's helping the shearer get the job done. When they do it right, shearing almost looks like a dance, a sort of ancient embrace between shepherd and sheep. You can really imagine people doing this work for thousands of years.

But we wouldn't just be watching the alpaca shearing. Gay and Mike made it clear that this was a hands-on, team event. They would fly up an expert alpaca shearer from Christchurch, pick up all their customers' alpacas, and bring them back to the central farm. They'd charge us a very reasonable price for the shearing, but in exchange, we had to volunteer an entire day's work. Apparently, each alpaca took four people to subdue. That was our first clue: this would not be an ancient embrace.

Gay and Mike picked the boys up on a Wednesday, while we were running errands in town. I don't know how they subdued them. It's possible they forced black hoods over their heads and hustled them into a waiting van, because when we got home that evening, the alpacas had disappeared. We took the opportunity to stroll peaceably through our paddock, admiring the lily pads in the pond and the tiny green frogs chirping on the bank. We'd never seen those details before. When our alpacas were around, we were always too busy dodging slime.

The big day was Sunday, and the three of us dropped the kids off with Autumn and Patrice before driving out to the alpaca farm with a creeping sense of dread. There were eighty alpacas to shear and only ten humans signed up for the day. If anything went wrong, we were dangerously outnumbered. But I tried to recall the sheep shearing and how peaceful the sheep looked when shorn.

We arrived at the farm with hopeful smiles, and Peter got his camera out to record the memories. Some man put his hand up. "No filming, bro, sorry," he said.

Peter lowered the camera. "No? Why not?"

"Eh. We've had some problems with the animal rights people." He picked up his shearing tools and abruptly turned his back.

Rebecca seemed nervous. "The sheep were very peaceful," I whispered. "I'm sure it'll be fine."

His tools assembled, the shearer turned back and introduced himself as Scott. Brisk and no-nonsense, Scott had a shaved head and a sharply trimmed goatee. He clapped his hands to get our attention.

"Right, you lot, this is how it goes. We bring the first beast out here. You and you"—he pointed to Peter and another man, a middle-aged guy fiddling with his smartphone—"are on the front legs. You secure the legs with these." He held up a pair of heavy ropes locked to a post in the ground with a steel carabiner. I recognized the lines, because we had them on our sailboat. You could use them to tow a large car.

"Then you and you"—he pointed to me and Rebecca—"you get the hind legs. On my mark, the four of you haul your lines, and I mean *hard out*. If you can't pull hard enough, put the bloody thing over your shoulder and walk away with it. That'll take the bugger down."

Shearing an alpaca was beginning to sound like subduing a dangerous psychopath. I shot Peter a look of concern.

The first animal they led out was a black alpaca named Odysseus. Despite his noble name, he had a scrappy look in his eye. Scott smiled grimly. "Know this bugger," he informed us. "He's a spitter."

Odysseus did not look at all pleased. We hauled on our lines, and his legs splayed out. Once Odysseus's face was level with mine, he shot a gob of green goo right at my head. I couldn't wipe it off without dropping a line, so I kept on pulling, feeling the warm slime slither through my hair and soak into my scalp. When Odysseus was subdued, Scott dove in and bound his head to his leg with a bungee cord.

I don't think this hurts the alpaca so much as insults his dignity. By this point the fleece is so thick and wooly that it must be a relief to have it removed. But that didn't stop Odysseus from complaining. He protested with a high-pitched scream that sounded a lot like a car's squeaky fan belt. The shearer straddled him hard, clenching him in powerful legs, and shaved fleece off his back and belly.

Odysseus just lay there and howled. At one point, Scott leaned in close so he could check the animal's teeth. Odysseus shot out a geyser of slime, tossing his head for maximum impact. Thrashing his head in a foul-smelling puddle, Odysseus looked like a belligerent drunk throwing up at the end of a bar fight.

The shearing was finished in a couple of minutes, and we released Odysseus unharmed, though clearly humiliated. Shorn of his magnificent fleece, he now resembled a skinny giraffe. He slunk back to his pen, chastened and nude.

We spent the morning shaving alpacas, watching them vomit and writhe, until Gay had us break for tea.

"You see?" she sang out. "Isn't it wonderful? They look so cute when they're shorn!"

Rebecca and Peter nodded politely, shoveling scones in their mouths so they wouldn't have to respond.

"Definitely," I told her. "They're great." To Peter, I whispered, "I've changed my mind."

He raised his eyebrows, his mouth full of scone.

"I don't want a pashmina. This whole thing is stressing me out."

Peter swallowed, taking my hand. "I'm so glad you said that. I'll buy you one myself. Let's get the hell out of here." Rebecca nodded vigorously, already reaching for her handwoven scarf.

Over the course of the morning, more farmers had arrived to help out, so we didn't feel too guilty about leaving early. I grabbed one last scone off the table. "Gay? Mike? We're gonna call it a day. See you later!"

The alpaca crew, oblivious to our discomfiture, waved a cheerful good-bye, promising to deliver our animals later in the week. We beat a hasty retreat.

"Maybe we're just too sensitive," I reasoned as we all drove home.

"Sometimes you have to *subdue* an animal to make it do what you want. Sometimes you have to squeeze its boob, or shave its ass, or—"

"Bag its face and strap it to the ground?" Peter asked. "Maybe we're not cut out to be farmers."

"You guys are doing fine!" Rebecca protested. "You're just learning. No one's an expert at first."

"See?" I backed her up. "Farm girl says we're doing fine. Who'd be more cut out for this than we are?"

"Axe murderers," Peter suggested. "Sex perverts. People who enjoy a nice goat vagina."

"Yeah," I conceded. "I see your point. But I wanted to leave early anyway. Now I have time to start a cheese."

This was getting more urgent now that Pearl was about to give birth. Soon we'd be flooded with gallons of raw goat milk, and I wanted to start practicing cheese making. I'd coaxed Hamish into giving me some fresh cow's milk—the kind without any dirt or snails in it—and I was thinking I'd start with a Camembert.

Cheese making is easy—or that's what I thought before I tried it. You just warm up a pot of milk and mix in germs to make it curdle. Then you squirt in a clear serum known as rennet, which is squeezed from the stomach lining of a dead baby cow. When the rotten milk coagulates, you hang it somewhere to drip like a suppurating wound.

And basically, that produces a cheese.

But if you're making an aged cheese, such as a Camembert or a blue, then the process continues. You might spritz your masterpiece with a bottle of live mold. Or inject it with infectious spores, as if your cheese were the victim of an anthrax attack.

Some of the best cheeses are the most disgusting, such as the French Mimolette, which is swarming with millions of live cheese mites. Then there's the Sardinian Casu Marzu, which is populated with maggots. These maggots, mind you, are alive. Sometimes, they jump out of the cheese and pelt you in the face.

If you are a fussy person who doesn't eat live maggots, you can slip your Casu Marzu into a small paper bag and shake it. This makes the maggots jump out of the cheese and hop around in the bag like popcorn.

But Casu Marzu is advanced, a lofty goal beyond my skill set. So I started with a simple Camembert. My first task was infecting the milk with a specific strain of bacteria. Not just any bacteria. I couldn't just cough in the milk or dip an old sock in the curds. That would have introduced bacteria, but it wouldn't be quite the right kind. Instead, I'd purchased a dime bag of *Penicillium candidum*, the mold that produces the snowy white surface found on Camembert and Brie.

I sprinkled the mold over lukewarm milk, covered the pot, and waited for the magic to happen. In just three hours the curds were set, and I had a substance that looked very much like a pale, watery Jell-O. I spooned the curds into cheese molds, where they glistened and dripped, looking nothing like anything that would ever be cheese.

The recipe said I had to turn my molds every hour, so I set my alarm throughout the night. Each time it rang, I staggered into the kitchen to flip the dreaded curd. Just before dawn I scrubbed out our spare refrigerator with bleach and hot, soapy water, then sprayed the walls down with vinegar to keep out invasive spores and encourage my cheesy ones to grow.

If this all sounds a little obsessive-compulsive, I may need to explain something about my childhood. For most people there's a particular smell they associate with the comforts of home, such as a berry pie baking in the oven or the scent of your mother's breast when she pulls you in for a hug. For me, that smell is rot. Specifically, the scent of comfort and love is the smell of a rancid cheese emerging from its sweaty wrapper.

My father loved food, and he held a particular fondness for the nastiest French cheeses he could buy. Reblochon, Roquefort, Morbier—the older, runnier, and more gag-inspiring, the greater his delight. He used to keep these decaying lumps in a separate drawer of our family refrigerator, and each night, after dinner, with lip-smacking anticipation, he'd pull out a selection of plastic-wrapped blight and spread these secretions on pieces of baguette. As a small child, I used to run away, refusing to come back until the cheese course had come to a close.

But as I grew up, I got a taste for it, and now the tangy scent of rancid cow's milk being slathered on a slice of crusty bread is enough to make my lips tingle and my mouth water. I don't know what happened. They say that certain opiates will make you throw up the first time you try them. Then, the next thing you know, you're hooked for life. That's pretty much how I feel about stinky cheese.

The problem is that most of New Zealand doesn't agree. To my dismay, Kiwis persist in the belief that cheese should be appetizing. Serve them a nice warm Morbier redolent of putrefying cow shit and heaven, and your average New Zealander will assume he's been poisoned.

So I had to make my own. For the next two weeks, I flipped my cheeses faithfully, monitoring the temperature and the humidity in their ripening chamber, misting them with cool water like overheated debutantes when they got a little parched.

But then I had to leave for a week, because my oldest friend, Ames, was getting married. Our mothers had been best friends for decades, and Ames and I used to crawl around together on the same baby blanket. Now he was running an online literary magazine in Brooklyn, and was marrying a Croatian-Colombian poet at his mother's home in Marin County. There was no way I could miss it.

But I couldn't bear to abandon my cheeses. I was happy to leave twenty animals and my own two kids with Rebecca and Peter, but the cheeses were another matter. "They're so vulnerable!" I opined, handing pages of detailed instructions to my niece. "They're only a few weeks old!"

"Go," Peter ordered. "Relax. Enjoy yourself." But I couldn't relax. I'd been out of the United States for years, and I hadn't lived in New York for more than a decade. This wedding party would be full of sophisticated city people, the women wearing real couture from boutiques I'd never heard of. Tucked away in my suitcase was a little black dress I'd purchased on clearance from one of Whangarei's few upscale shops. My black boots were discount. Though technically I was flying home, I felt like Country Mouse taking on the Big Smoke.

So when I got there, I was startled to learn that Peter and I were apparently living the Great Brooklyn Farm Dream. Ames and Danica's wedding guests were a circus troupe of New York creativity, everyone with glamorous-sounding jobs. They were writers, directors, musicians, and puppeteers. "It premiered at Sundance," was a phrase I actually overheard among the cocktails, along with "We were so excited when Oprah chose it for her book club."

But despite all the glitter, the theme of the wedding was country farm chic. The invitations showed Ames and Danica wearing plaid and posing together in a field full of wheat. When I arrived at the house, a graphic artist was carefully stenciling the bar menu in chalk on a blackboard, and the nervous bride put me to work wrapping potted plants in burlap. Ames announced to his friends that I lived "on a *farm* in *New Zealand*," and one woman actually squealed, sharing aspirational tales of backyard chickens and urban beehives.

"How long are you out here?" she asked, sipping an artisanal bourbon.

"Just a week," I told her. "I have to get back. My goat's giving birth any day now."

"Oh, my God," she giggled. "That's too funny."

"No, I'm not kidding," I told her. "I have to get back. For the . . . kidding."

"Can you help with the seating?" Ames pressed my arm, leading me to a grassy area by the swimming pool. "We're gonna have the guests sit on hay bales, to make it more farmy."

"Great," I said, grinning. "But that's not hay."

Danica overheard me. "It's not? *Shit.* I *knew* this was going too well. What is it?"

I kicked at the bale. "This is straw. It's better for seating, I guess. It's cheaper than hay. But you can't feed it to animals. It doesn't have any nutritional content." This little treatise on animal feed came out before I realized what I was saying.

"Sorry," I blurted, embarrassed. "That . . . was probably more than you needed to know."

Ames and Danica stared at me. "Oh, my God." Danica burst out laughing. "We are such *idiots*! How are we ever gonna get our own farm if we don't know the difference between hay and straw?"

Ames just shook his head and slapped me on the back. "That's awesome." He chuckled. "Look at you! You're a country girl! Now move this fucking straw so I can get married."

The wedding went smoothly, and the ceremony was short—a relief, since straw bales don't make the most comfortable chairs. But I didn't want to stay up all night quaffing mojitos and dancing by the swimming pool. I was worried about my goat. So I slipped out early to call my family on Skype.

"Has Pearl given birth yet?" I demanded as soon as they picked up.

"Not yet!" Rebecca chirped. "But it's happening any day now."

"Don't worry," Peter soothed. "Have fun. She's fine. We're all fine."

But he was wrong. Everyone was not fine. At the end of the week, I boarded my plane to Auckland, hoping to sleep through the night on the long transpacific flight. I pulled on a warm pair of socks and flipped idly through the selection of Hollywood movies. I ate bland chicken with a plastic fork and held up my cup when the flight attendant offered a second glass of wine. I woke up as we began our descent, and before long we were sliding to a stop in New Zealand.

When I turned on my phone, I found a text message from Peter. "Call me on my mobile," he said. "It's important."

Still fuzzy from my fitful sleep, I glanced at the time on my phone: 6:00 a.m. *Why would I call him on his cell phone? He hasn't left for work yet. He must be confused.*

So I called our home line instead. The phone just rang and rang. Then the answering machine picked up.

And that's when it hit me. If Peter wanted me to call him on his cell phone, then he was in town. And there was only one reason he'd be in town before dawn on a workday.

Someone was in the hospital.

CHAPTER TWELVE

FROG IN MOUTH

Heat rising in my cheeks, I called the cell phone. Peter picked up on the second ring.

"Baby?" he said. "We're okay. Silas had a seizure."

Maybe it's from years of sailing together, but Peter and I almost always speak calmly to each other, even in a crisis. We've been in some dangerous spots, when a single bad choice could have crushed the boat. Even sailing over treacherous sandbars or threading the narrow entrance to a reef, Peter never raised his voice. He always sounded clear and calm.

I, on the other hand, was not calm. "What does that mean?" I asked, my throat clenched. "What kind of a seizure?"

"A big one. A . . . grand mal. Then he had a—whoops, there he goes again. Sorry, baby. Gotta go."

The line went dead. I stared at my phone. The plane was mostly empty, and flight attendants were moving down the aisles, collecting trash and forgotten magazines.

"Ma'am?" A pretty young woman dressed in the jade green Air New Zealand uniform approached me. "Did you need some help today?"

"No, I don't know. My—son just had a seizure." I got to my feet and realized I was still in my socks. I looked dumbly at the flight attendant. "And I can't find my shoes."

I was waiting in line at customs when Peter called back to explain. Twenty hours prior, as I was dropping off my rental car and getting ready to board my flight, Silas had a grand mal seizure in the back of the school bus.

"Rebecca got him to the hospital," Peter explained. "Amanda came, too. I met them there, and the doctors were just trying to decide whether or not to admit him when he started having another seizure, right there in the emergency room. So they kept him overnight. I'm with him now."

"Where's Miranda?" I demanded. "Is she okay?"

"She's fine. Rebecca took her to day care. Then, this morning, when you called, he had another one. That one was pretty bad."

Peter has spent a great deal of time on the ocean by himself, and he's not one to exaggerate for dramatic effect. So when he said "pretty bad," my stomach flipped. I hung up to get through customs, then I had to board a commuter flight to Whangarei.

By midmorning my last plane had landed. Rebecca was at the airport to meet me. She hugged me without saying anything, her eyes filling up with tears. I didn't want to notice. There wasn't any time for that.

On the car ride to the hospital, Rebecca explained what had happened the day before. She'd walked Silas out to the school bus, as she'd done every day that week. She clipped him into his seat belt and waved good-bye.

"Bus," said Silas. "Bus-ah-bus." Then, strangely, he said, "home."

"You've been at home," Lish sang out from the driver's seat. "Now it's time for *school*, Silas."

Rebecca slid the door shut and strolled back down the driveway, getting ready to feed the chickens and start on the breakfast dishes.

Lish said she'd heard something tapping behind her seat. She glanced in the rearview mirror to see Silas stiff, his arms and legs straight out like a robot's, his eyes rolling back in his head. He had a little blue Smurf toy clutched in his fist, and he was hammering it against the window. His face went gray. Foam started bubbling out of the side of his mouth.

She slammed on the brakes, unclipped her seat belt, and switched off the ignition, diving into the backseat. She grabbed Silas by the shoulders and eased him to her lap, releasing his seat belt, holding him on his side while he shuddered and shook.

Two other first-graders were riding with them that day. "*Why* are you holding Silas?" Ethan demanded.

Silas was still in midseizure, his eyes rolling, and Lish didn't want to scare the others. "He just has a little froth in his mouth, Ethan," she explained.

"*Why* does he have a frog in his mouth?" Ethan wanted to know, and then Silas's shaking slowed, went rhythmic, as though electric shock waves were pulsing through his body. He blinked. The color started coming back to his face.

Lish held him in her arms, rocking him gently. After a minute or two his eyes began to focus, and she said, "Where ya been, poppet? You were a million miles away."

Rebecca went on with her story. "Then the school bus came back to the house, and I thought, *What happened? Did I forget Silas's bag?* And then Lish carried him out and I just—I didn't know what to do."

"Was he still seizing?"

"No, he was—really asleep. Lish had to take the other kids to school, so I called Amanda, and Amanda drove us to the hospital. I met Uncle Peter there."

I glanced at her. "Sounds like you did all the right things." Rebecca was a sheltered twenty-year-old kid who spent most of her time spinning yarn, but she'd managed a crisis with a five-year-old in a foreign country. I was impressed.

We pulled into the parking lot of Whangarei Hospital, a vast public building of bleak white concrete. *Here we go again*, I thought. This was the hospital where they treated Silas when he was first born, because he was weak and needed support. It's where the psychologists told us he was "delayed," because no one likes to say *retarded* or *handicapped*. It's where we went when he tumbled off a chair and broke his arm, so they could shoot him full of morphine and set his bones with plaster and steel.

Each new crisis with Silas takes us back to the hospital. And now here we were again, rushing to the children's ward because Silas's brain had gone offline with no warning.

Peter was sitting by Silas's bed, halfheartedly thumbing through a paperback. Silas was tucked under clean white sheets, sleeping deeply. A thick white bandage was tied around one arm.

Peter hugged me, and then I pointed to the bandage. "What's that about? Did he hurt himself?"

"They put a line in, just in case they want to give him something intravenously. He kept ripping it off, so they bandaged it up."

A hard plastic chair stood at one end of the bed, and I collapsed into it. All I could think was *Brain tumor*. Finally, I had to say it out loud.

"Why else would a person suddenly start having seizures?" I

hissed. "His brain just stopped. There's got to be something *growing* in his *brain*."

Peter shook his head. "I don't know. The pediatrician's coming to talk to us soon. You could ask him about it." Restless, I pulled out my phone and Googled "epilepsy brain tumor."

If there's one thing you don't want to Google when you're worried about your child's health, it's "epilepsy brain tumor." I read about glioblastoma tumors, evil tentacles that ensnare your kid's brain and squeeze until there's nothing left. Swallowing hard, I shut off my phone.

Silas's pediatrician is Dr. Osei, a tall, elegant man from Ghana. He strode in later that morning holding Silas's thick case folder under one arm. "Well, hello," he said, greeting me. Dr. Osei's speech has a slightly exotic cadence, his vowel sounds especially round. "I thought you were on holiday?"

"I got in this morning," I explained. "I came right here. Do you know what's happening? Why is Silas having seizures all of a sudden?"

The doctor raised his eyebrows, a humorless smile at his lips. "That is the big question, of course. And at this stage we do not have an answer. As I explained to your husband, every child is allowed one fit. But now Silas has had several . . ." He trailed off, looking down at Silas's sleeping form. "It's difficult to know why this would happen," he explained. "Seizures are quite common, with . . . these sorts of children."

Peter and I exchanged a look. *What sorts of children?* I wanted to snap. *The kind who listen to Broadway musicals? The kind who love hopping up and down and messing with their father's computer?* Of course I knew what he meant: handicapped children, children with intellectual disabilities. But I didn't ask him to clarify.

Instead, I took a deep breath. "Is it possible he has a . . . brain tumor?" I asked. "I mean, how can this just start, out of nowhere? How come my son is epileptic all of a sudden?"

I looked at him as if he were some kind of oracle, but Dr. Osei just cleared his throat. "I can tell you that it's likely not a brain tumor. Localized tumors produce asymmetrical seizures. You might see one or two limbs going stiff, but not the whole body. As to why, it is difficult to say. We will investigate, of course. But we might never know."

I'd heard that one before. "Okay," I conceded, shifting on my plastic chair. "But there're drugs for this, right? I mean, we'll get the seizures under control, and then we won't have to come back to the hospital?"

"Likely not." The doctor glanced at his chart. "There are a few red flags. If he gets into a cycle of seizures, as we've seen here, yesterday and today. If he has another seizure, without waking up from the first one. And"—he scribbled something on a prescription pad and handed me the paper—"if he has a seizure that lasts more than five minutes, you will need to give him this medicine. It's Midazolam. Then you must call an ambulance."

"Why?" I wanted to know. "What happens in five minutes?"

Dr. Osei smiled apologetically. "A slight possibility of brain damage. You would not want it to go that long."

Silas was discharged that evening, and we all drove home. We had two cars at the hospital, so Peter and Rebecca took the kids home in one, and I stopped off at the pharmacy. I had two prescriptions to fill, the emergency medicine and a cherry-flavored syrup they'd said would control the seizures. Standing in line waiting for the drugs, I thought I should feel sad or angry, but I didn't. I felt punched in the gut.

Halfway back to the house I started breathing too fast, and I wasn't sure if I was having a panic attack—until I pulled to the side of the road and wept. Nothing we did made a difference. No one knew why Silas was delayed, or why he couldn't talk, or why his brain was now flickering in and out. And if no one could figure out what was wrong, then nothing could ever be fixed.

Peter and the kids were home when I got there, the front deck lit up from the living room. Through the sliding glass doors, I could see Peter easing Silas into clean clothes. The spinning wheel stood in the middle of the living room floor, looking reproachful. Three large sacks of alpaca fleece leaned against the wall, waiting to be spun into yarn. I felt like that nameless sucker, the miller's daughter in "Rumpelstiltskin," the one who needed a miracle but didn't have a clue. Who has the skills to spin straw into gold?

"Hey," I said, coming inside. "How's the patient? I scored his drugs in town."

"Seems good." Peter tucked him up on the couch with his Dart. "Tired. Rebecca already went to bed."

Miranda started tugging on my shirt right away. "Mama," she announced. "I would like one chicken bone for my dinner please, and a chunk of cucumber. And two sprinkle cookies for dessert."

I glanced at Silas. All I wanted to do was lie down with him, curl my body around his and sleep. But Miranda still needed to eat, so I opened the refrigerator and peeked inside. A flabby pink chicken that would have to be roasted. Lettuce that had to be washed. Suddenly, it all looked impossible.

"Hey, there's a message on the machine," Peter commented. "Did you check it?"

I pushed the Play button. "Just wondering how you guys are." It was Amanda's voice, sounding like a fairy godmother's on the scratchy phone recording. "Thought you might not feel like coping with dinner. We're having something simple, if you want to come round."

She didn't need to ask twice. When we got to her house, Amanda was pulling two roasted chickens from the oven, the steel pan dripping with golden juice. A huge tray of potatoes, pumpkin, and kumara waited on the dining table. She poured oversize glasses of white wine and passed steaming plates of food. Seated around her dining table with her three daughters all chattering about spies and spooky ghosts, I felt absurdly grateful.

Still exhausted from the hospital, Silas curled up on the couch. Miranda leaned in next to me, contentedly gnawing a drumstick. "Do you understand what a seizure is?" I asked her when there was a lull in the conversation.

"Yes." She nodded solemnly. "Silas has a sneejur because his brain doesn't work so well."

"I guess you could say that."

"And he can't talk like me."

I shook my head. "No, he can't."

She sneezed and wiped her nose on my blouse. "Mama, I think I just had a sneejur."

"No, that was a sneeze."

"And you have to take me to the hostibal."

"I don't think so."

"And if a crocodile tries to bite me, then you will take a gun and dead him."

Amanda's husband, Nick, sat on one end of the table, tucking into his chicken. "Bit worried about the seizures, are we?" he asked.

"I guess we all are," I admitted.

"It's quite common," he pointed out. "Lots of the kids I work with have them. Sometimes dozens of fits a day. Big ones, little ones. Some of them have seizures all day long."

This wasn't exactly comforting. "I think maybe he's been having them for a while," I observed. "All that heavy breathing, losing his words the past few weeks, until all he could say was 'bus.' Do you think maybe he's been having mini-seizures, and no one ever noticed?"

"Could be." Nick nodded thoughtfully. "I really don't know."

If I heard that phrase one more time, I thought I might start throwing things. I decided to change the subject. "Skin was over the other day."

Amanda raised her eyebrows. "Oh, yeah? Teach you how to butcher a pig, did he?"

"No, we finally killed our boring turkeys. Then we plucked out the feathers and opened them up, and it turned out they were full of rocks."

Nick blinked. "They eat rocks? I knew they weren't that bright, but . . ."

"No, it's on purpose. They hold the rocks in their gizzards, and it helps them to grind up their food. They don't even have teeth, but the rocks are like teeth on the inside."

Amanda put down her fork. "That's life on the farm, really," she said. "Never know what you'll find down there. Pigs running off, rooster killing . . ."

"Rooster killing?" Peter interjected. "We could use some of that."

"Nick had to kill our rooster." Amanda nodded proudly at her husband. "They all get mean at some stage, then they attack the kids. Or they gang-rape their mother."

"Excuse me?" I looked up from my plate. "You didn't mention that when you gave us Goldie."

"You didn't ask." Amanda smiled wickedly. "Anyway, you were so keen for a rooster I thought I'd let you find out for yourself."

She had a point. I had asked for the rooster.

Amanda sipped her wine. "When I was a girl, Mum thought it would be good for us kids to get a chicken with a clutch of eggs, so we could see the baby chicks being born. And it was nice. They all hatched out of their little eggs, and they were really cute."

"So what happened?" Peter asked.

"They all turned out to be roosters. Every last one of them. Then they grew up and gang-raped the mum."

"Oh my *God*."

"What's a gang rape?" Miranda wanted to know. "Is that a grown-up word?"

I shushed her, and Amanda giggled. "I know. It's shocking, really. We were having a teddy bear birthday party, if you can believe it, and all my little friends were over, with our teddy bears and cake, and the dog got out. And she deposited this—er, this *well-loved* chicken in the middle of the picnic. Dead."

There was a silence as we took this in. Then I burst out laughing.

Peter shot me a reproachful look. "How can you *laugh*?" he asked. "It's chicken matricide and sexual assault!"

"It's just . . ." I took my napkin and dabbed my eyes, helpless with giggles. "It's so fucking hard to be a parent. And no matter how bad it gets, it's always so much *worse* for the animals."

"I don't know," Nick objected. "Our animals have it pretty good. You should see how they keep pigs in Malaysia."

"Oh, I'm sure you're right," I said. "But we're so stressed out about Silas's seizures, and I have to say, no one's getting raped by a duck. Or getting his nose hacked off with a laser beam. Or getting infested with thousands of vampire worms."

Sophie, Lucy, and Amelia were all staring at me, their dark eyes round with horror.

"Okay!" Amanda got up from the table. "That's enough nightmares for now. Anyone for a pud?" She dished out bowls of warm chocolate pudding, passed a pitcher of cream, and sat down to eat. "Every parent goes through some version of this. We've had our share of surgeries. The allergies, the infections. It's always awful."

"But they said if it's more than five minutes, we have to call an ambulance," I protested. "We have this emergency medicine, and—"

"If Amelia eats a peanut, *we* have to call an ambulance," Amanda interrupted. She gave me a hard look. "One nut."

"Have to eat the rocks with the maize," Nick observed. "Get some of those inside teeth growing."

The doctors told us to expect more seizures, and for the next few days we watched Silas closely. He couldn't be left alone now, especially not in the bath. The neurologist was still adjusting his meds, and every few days, the seizures came and took him away. He'd be standing in the middle of the living room, turning the spinning wheel, and I'd hear the familiar chant:

"Bus . . . ah . . . mmmm." Then I'd glance his way, only to see an ecstatic smile take over his face, his eyes rolling back in his head. I'd lunge across the room as he crumpled to the ground, and sometimes I'd catch him before he fell.

The fits came in waves, a series of maritime squalls, small storm fronts that blew in without warning or cause. When Peter and I were sailing, we traveled through some tricky weather. Motoring across the Gulf of Panama during storm season, I'd look on the radar screen and see them: multiple clusters of black dots, patches of high wind and rain so dense they'd show up as solid masses.

There were too many to dodge, so we let them blow over us, shake our rigging, and pitch us sideways. They'd disperse in a minute or two, leaving us in the eerie silence of the Doldrums.

Peter and I loved that part of sailing. We couldn't insulate ourselves from the weather, the way people did in their large, heated homes in the city. On our boat, we felt connected to the rhythms and whims of the planet. When we knew a storm was coming we shortened sail and battened the hatches, sat back and enjoyed the lightning, confident in the strength of our vessel.

But sailboats are built to weather storms. Each time one hit, we emerged from the squall unharmed. That's not how it was with Silas. He'd be seizure-free for a few days, even a week. Then the storms would hit and knock him back down. When he came to, he'd be confused, and his words would disappear. "Bus," he'd say, "bus-ah-bus," and all I could do was hold him.

Eventually it seemed silly to keep him home all day. I couldn't stop the seizures just by cuddling him and reading him storybooks. We gave his teachers specific instructions about what to do if Silas went down, sent him back to school, and hoped for the best.

So on Pearl's big day, Rebecca and I were home alone. We'd packed off Silas and Miranda to school, and Peter had left for work. I was pottering around the farm, ostensibly feeding animals but really looking for reasons not to write. I checked Pearl's vagina, and that's when I saw it. *There were wood shavings stuck to her butt.* This might not seem like an extraordinary discovery, but it meant there was something sticky back there. I looked closer.

"Hot damn." I breathed. A thin stream of mucus hung between her legs. "Rebecca!" I called. "Come and see this!"

She ran over, and squeaked when she saw the mucus. The two of us clapped and stroked Pearl on the neck, saying inane and encouraging things such as "Good job, Pearlie Whirlie!" And she even let us touch her. It was possible she'd realized she might need us soon.

A couple of hours later, I was hanging laundry out to dry when I heard the demon death knell. "GWAAAAAK!" came a noise unlike any sound my animals make. "What was that?" I called, rushing to the front of the house.

Rebecca was right ahead of me. "Pearl!" she gasped, and then she got to Pearl's side. "She—uh—ah, there's a baby coming out!"

Pearl was lying on her side beneath the palm tree. And a goat head was coming out of her ass.

Not her ass, exactly, because that wouldn't make sense. But as I may have mentioned, it's all kind of a mystery down there, and there was definitely a goat head coming out of a hole in her hindquarters. Pearl was amazingly cool for someone who was shitting a live goat out her vagina, as though she'd done this every day since the beginning of time.

I took my cue from her. "Good job, Pearlie!" I coaxed. "Just one more push!" To Rebecca, I hissed, "What do we do now?"

"Hopefully, nothing," Rebecca whispered. "Unless there's a problem, it's best not to touch her."

Pearl's body tensed again, but this time she didn't cry out. And there, on the ground, was a wet and goopy goat baby.

Rebecca squealed. I squealed. There was a lot of squealing that day. To be honest, I felt a little teary-eyed and sentimental. I'd never actually attended a birth where I wasn't personally pushing out a melon, and it was easier to appreciate the Miracle of Life when I wasn't convinced I was dying. And it really did seem like a miracle. I watched in amazement as the baby unfurled its long legs and shakily got to its feet. It fell right away but didn't give up. It leaned its weight forward and tried again, and all the while Pearl was licking and cleaning it, swallowing the amber membrane as she worked. This little kid was a patchwork of black and brown spots, with a snowy white head and a black backside. It breathed right away, and once it managed to stand, it started sniffing at Pearl's armpit for food.

"Not there, honey." I nudged it. "In back." The kid was so eager to start living that I decided its name was Moxie.

"Is it a girl or a boy?" I asked Rebecca. "Can you tell?" The gender of this baby was critical. If it was a girl, we'd keep her for milk and cheese. If it was a boy, we'd be eating goat curry.

"I can't see yet; she has to pee." The little goat bent slightly at the hindquarters, urinating daintily into the grass. "It's a girl!" Rebecca crowed. "She pees just like Pearl, see? I think it's a girl!"

The pink goop dripping out of Pearl's backside was fascinating. I was filming it with my iPad, crouched at her hindquarters, and that's when I got the money shot. All at once, she blew a water balloon out her vagina. About the size of a grapefruit, it was full of amber fluid. She walked around like that unconcerned, a big yellow bubble hanging out her butt, and then I saw it: another snout, another hoof.

"There's another baby!" I hollered, as though I'd made this one myself.

The second kid came out in one push, surging to the ground in a flood of fluid. It was covered with membrane, its nose a bit more blocked than Moxie's. It spluttered a little, and Pearl cleared its face. This one, also a female, was pure white, with a black stripe down the spine just like her mother. Her name, of course, was Stripe.

The rest of the day was pure magic. Rebecca and I sat with Pearl and her babies, tweeting pictures of the birth and stroking the kids.

We made sure they could nurse, and once we figured out what the iodine was for, Rebecca held them up as I disinfected their little umbilical cords. We gave Pearl some hay, then rolled oats and raisins, then water that I'd sweetened with molasses. She ate all her treats, then she ate her placenta, then she ate the meconium right off her babies' butts, and then we thought we might be done with the magic.

"Thank you for the babies, Pearl," I murmured, scratching my goat behind the ears. "What a good job you did. Thank you for being so healthy."

She looked up at me then, her strange rectangle eyes full of light and intelligence. "And *thank you*," I continued, "for not making me insert my hand."

CHAPTER THIRTEEN

THE GRAB-AND-YANK

"My colleague Premal says we should fry it up. It's a delicacy in Sri Lanka," Peter announced when he got home from work. He was standing with me under the palm tree, watching Pearl nurse her babies. Moxie and Stripe were disturbingly good at it: their technique involved jumping up and smashing their mother in the udder to make her let down, a trick that looked painful and a little bit mean.

"Colostrum?! We can't eat the *colostrum*," Rebecca protested. "It's not even real milk. It's like a vitamin soup. It's got all the antibodies the babies need to survive."

"Whatever." Peter shrugged, then headed inside to change. "But Premal says it's the best stuff there is. He says you'll never put anything more delicious in your mouth."

Seriously? I sat there for a while longer, watching the babies prance around Pearl. *The best stuff there is?* How could we let this chance slide? When would I get another opportunity to taste pudding made from goat colostrum squeezed fresh from the teat?

"Mama?" Miranda came up behind me carrying a plastic baseball bat. "Do you want to come see this dead thing that I don't know what it is?"

"Of course I do, Magnolia," I said, rising to my feet.

The dead thing turned out to be a desiccated songbird, and we buried it in the compost before heading back inside. But despite my best intentions, I couldn't stop thinking about goat colostrum. That night I got on Google and learned that eating this stuff is all the rage. Just about everywhere people raise dairy animals, there's a tradition of tasting the mother's first milk. In Iceland, there's a pudding called *broddur*, and the French make a savory custard called *béton*. The English call animal colostrum beestings, and with that discovery, I made up my mind. There was no way I was passing on a fresh beestings pudding.

But first, I had to learn how to milk. Autumn had already loaned us her milking stand, a sturdy wooden contraption that looked somewhat like medieval stocks. There was a platform for Pearl to stand on, and a shelf where we put her food. Once she'd jumped on the platform and bent to eat her breakfast, we locked in her head with a sliding plank. She was trapped, but this didn't seem to bother her. As long as she had something to munch on, Pearl was happy to be fondled.

This was a subject Rebecca felt confident to teach. She'd milked her way through high school and had taught countless students how to do it, usually in a freezing barn before dawn.

"Sit there," she instructed, pointing at the seat that jutted out from the milking stand. "And grab a teat in one hand."

I did as I was told. Pearl jerked a little when I grabbed her boob, but she didn't run away. Possibly this was because her head was locked in place.

Rebecca bent close. "Now, gently but firmly, ease the pressure down to the end of the teat."

I did. No milk.

"Too gentle. Try using a little more pressure."

I squeezed the teat hard, and Pearl squawked in alarm, kicking her hind leg at my face.

"*Not that hard.* See?" Rebecca took over and showed me how it was done. Contrary to what I'd thought, milking an animal isn't just a grab-and-yank. You increase the pressure as you move downward, coaxing the milk to the tip of the teat.

I think I must have exhausted Rebecca, because after our lesson, she retired to her sleep-out.

"It's a lot like giving a hand job," I reported to Peter, once my lesson was finished. "You go easy at first and see what you get."

"I never liked hand jobs," Peter observed.

I winced, remembering the grab-and-yank I'd performed on poor Pearl. "You never know," I said brightly. "Maybe milking will improve my technique."

"So is it hard?" he asked. "The milking?"

"Not once you get the hang of it," I bragged. "At first it takes a little while. You kind of have to stimulate her, to get her going."

Peter put down his coffee. "What do you mean, you punch her in the tit?"

"Kind of," I turned the stove on, readying my pudding ingredients. "Not 'punch her in the tit' so much as massage it. You should see how those babies go at her. They're relentless."

As if on cue, Miranda walked into the room. Still groggy from sleep, she stood there in her pink rabbit nightgown. "Mama?" she asked.

I was stirring my colostrum.

"*Mama,*" she repeated.

I put some butter in the pan and watched it melt.

"MAMA!" she shrieked, as if someone were breaking her leg.

"*What?*" I whirled around.

"Can I have a cuddle?"

"Can't you see I'm making pudding?" I snapped, sloshing the colostrum in the pan, where it instantly started curdling.

"Bus," Silas announced, entering the kitchen with an empty jar of mayonnaise.

"Has he been in the recycling again?" Peter got to his feet. "Silas, that's *garbage.* Come on, let's throw it away."

Ignoring his father, Silas bent to the floor and started spinning the empty jar.

"No one is giving me a cuddle!" Miranda raged. "Not even one person!"

I looked skeptically at the colostrum, which had congealed in the pan like pale scrambled eggs. "Breakfast time!" I sang out. "I have something special for you today!"

The children eyeballed their bowls of goat beestings warily, and Peter got up from the table. "I'm off to check the cows," he said, and I turned back to the project at hand.

The second time I heated Pearl's colostrum, it was much more successful. This time, I heated it gently, over a double boiler. And I'll be damned if it didn't set up, turning thick and silky like a rich, eggy custard. I put a spoonful to my mouth, and I had to admit it was delicious. I could imagine it sweetened with saffron and cardamom, an exotic treat from the Orient.

But I never had time to bust out the cardamom, because Peter started hollering from outside.

"Antonia? Why is there blood all over the deck?"

I turned off the stove and rushed outside, only to find a pattern of blood drops speckling the deck, from the front door to the barbecue, then down the two front steps.

"Miranda?" I called. "Are you bleeding?"

"Mama, this custard is weird," she replied. "I hate it."

I ran inside to inspect Silas, but he was intact, too. "Do you have an ouchie?" I asked, knowing he wouldn't be able to answer. Still, I lifted his Bob Marley T-shirt and felt around for any injuries. "Bus," Silas replied. "Bus-ah-bus-ah-bus."

"*Miranda*," I insisted. "Eat your pudding. It's full of antibodies so you don't get sick."

"But I'm *not* sick," she whined.

"Still," I told her. "It'll protect you from hoof rot. And black leg. And mange."

Miranda rolled her eyes and pushed past me, her colostrum pudding untouched. "I'm gonna go see my baby goats," she announced, and since I was busy solving the mystery of the blood spots, I didn't argue.

My next thought was that Kowhai had killed a chicken and dragged its hapless body over our porch before hiding in the bushes to devour it, so I stood at the door and yelled. "*Kowhai! Come. Here. Now!*"

Peter was down on his hands and knees, wiping bloodstains off the deck, when Kowhai came trotting up. "You look guilty," I said when I saw her. "Who did you kill?"

"Oh, Jesus." Peter cringed.

"What?"

He leaned back on his knees and pointed. "The dog has her period."

I turned Kowhai around and examined her. Sure enough, little droplets of blood were clustered around her backside. We knew this would happen, since we'd planned to let her have a litter. Still, it was a little early in the day for dog period.

Peter got to his feet. "Now I'll have to build a bitch box." He sighed. "Where the hell am I gonna put one of those?"

That hurt my feelings. "Seriously? I know I don't give the best hand jobs, but—"

"Not for *you*." Peter snorted. "For the *dog*. Now she's in heat, we'll get male dogs sniffing around from all over, and they're all gonna want to mate with her. We have to lock her up so she can't get out."

"And it's called a bitch box?" I asked doubtfully. "That seems unkind."

"*Mama!*" Miranda called from the palm tree where she was hanging out with the goats. "Pearl is all bloody!"

"*Christ.*" I rolled my eyes. "Does this ever *end*?" It occurred to me that there was a reason Rebecca preferred her nail bed to hanging out with the rest of us.

Pearl's backside was crusted with dried black blood, so I got out a rag and a bucket of warm water. I wasn't worried about a little bleeding after the birth, but it was unpleasant to look at. Also, it occurred to me that flies might lay eggs back there, and I didn't want to deal with the maggots.

I locked Pearl into her milking stand and, as gently as possible, pressed a warm rag to her hindquarters, softening the bloody chunks and then picking them out of her fur. Mostly this didn't bother me, but when I flicked a bloody period chunk and Kowhai caught it in her mouth I must have groaned out loud.

"Oh, *God*." I moaned. "I could be in law school right now. Why didn't I go to law school?"

Miranda put her hand on her little hip and frowned. "Mama, you're just freaking out. What *next*?"

"I don't know," I admitted, wiping my bloody hands on my nightgown. "Certainly not law school. That ship has sailed."

I sat down on the milking stand, which Moxie and Stripe took as an invitation to play. Instantly, they were both on my lap, their hooves on my shoulders. Stripe started sucking my hair.

"No, Stripe, cut it out!" I sputtered, batting her away with the back of my hand. "My hair is *not* a breast!"

Bored with not being the center of attention, Miranda ran back inside. I returned my focus to Pearl's large, speckled teats. If I leaned my body in close to hers, resting my head against her side, Stripe and Moxie didn't have room to wiggle in between us. I pressed up against her, and Pearl sighed. Munching her nuts, she seemed to relax.

And that's when Miranda came running. "Mama!" she called. "We had a little problem with Silas! He had a seizure!"

I dropped the milk jar and ran, meeting Peter as he struggled out of the bathroom with Silas in his arms. Silas's body was stiff and gray, a thin child's corpse dripping water.

"Is he okay?" I asked, my voice too loud.

Peter laid our son on the living room carpet. "Can you get a towel?"

I ran for a towel, and we dried him off, holding Silas on his side as he shuddered and shook.

"Check the time," Peter barked.

"Eight forty-six." Silas was jiggling uncontrollably. The color was gone from his face. A thin stream of dribble ran down the side of his mouth.

Miranda stood behind me clutching That Baby to her chest. "Mama?" she asked. "Is Silas going to die?"

"No," I snapped.

"Miranda, be quiet," Peter ordered. "We need to concentrate." Something in his voice meant business, and for the first time in her young life, Miranda said nothing at all. She backed onto the couch and sat, burying her face in her doll.

"Time?" Peter asked.

"Eight forty-eight." Silas was jerking now, punching out his arms and legs, controlled by an invisible defibrillator.

"Forty-nine," I told Peter. "Do I go for the Midazolam?"

"Not yet." Then: "How long's it been?"

"Four minutes." I glanced over at the kitchen cupboard, where the emergency meds were kept.

"He's coming out of it." Silas's body was relaxing, but his eyes were still closed. Peter bent down to put his face at eye level with his son's. "Silas? You there?"

Silas did not respond. "Is he out of it?" I wasn't so sure.

Peter lifted Silas's eyelid. The boy blinked. With a yawn, he snuggled into his father.

"Bus," he murmured. I'd never been so glad to hear anything in my life.

"I was sitting right there," Peter explained. "He got some of that pudding down his front, so I was giving him a bath. I was just watching him splash in the tub. Then his eyes rolled up, and he slid right in."

"Did he go under?"

"Almost. I grabbed him just before. That would have been it," he said softly. "If I hadn't been there. He would have been gone."

I called the pediatrician; I wrote to the neurologist. "Why is this happening? When can we do an EEG? An MRI? Why is my son still having seizures when he's taking medication? What can you do to help?"

The answers were always depressingly vague. "In these kinds of cases, we manage the symptoms," the doctors told me. "The most important thing is to get his seizures under control." But they weren't under control. Every few days, Sophia would send Silas home from school because he'd collapsed at the squash court, on the playground, in the bus. Patrice was usually the one who caught him, and I could tell it was taking a toll. He'd retired from cooking to avoid heavy stress, and now he was catching an epileptic five-year-old for a living.

"Man, it is fucking scary," he told me one day when I'd arrived to pick up my son.

I nodded. "Tell me about it, Patrice." As usual after a fit, Silas was deeply asleep, so I scooped him up like I did when he was a baby and carried him to the car.

"Why aren't these doctors doing anything?" I raged at Peter. "It's like they don't even care, like they've written him off as some retarded kid who's just going to have these fits. Like it's normal."

Peter considered this. "We just have to push. We keep pushing until we get the help he needs."

"And what if we get kicked out of school? What if they decide they don't have the support for him in mainstream school and they make us send him to special school?"

That was my nightmare. Sending Silas to special school felt like giving up. As long as he was mainstreamed, there was some hope that he might become literate, learn to do basic arithmetic, function in society. Some of those kids in the special school were twelve years old and still learning how to feed themselves. Sending Silas to that place felt as good as declaring, "Oh well. Nice try. You'll be institutionalized for life."

Perversely, all my worry for Silas made me more involved on the farm. Rebecca and I split up the chores, working together to feed the hungry animals. I gathered the eggs, and she made sure everyone had fresh water. Not even Jabberwocky or the flesh-eating alpacas could frighten me now. I'd seen my son's eyes empty as his body shook with electric shocks. Compared to that, an angry rooster was a breeze.

November is spring planting season in New Zealand, and Autumn asked if she could seed our top paddock with corn. "It's a fantastic business idea," she said, grinning merrily at Peter. "We'll get early corn, then we'll sell it all for Christmas."

It sounded fine to us, and while Autumn and Patrice toiled in the upper paddock, I devoted myself to my goat cheese. Even though she was nursing her own kids, Pearl still gave us about a quart of fresh goat milk each day. I had grand plans to start a cheese business, a creamery based on milk from our own herd of goats, who fed on the local bush and made cheese flavored with our local *terroir*.

And yet goat's milk cheese seemed trickier than cow's milk. My first attempts at Camembert were edible, but Pearl's curds never turned into goat cheese at all. Instead, I produced a sort of tepid milk vomit, a curdled slurry that felt thick and lumpy but never firm.

"Maybe there's a market for that," Peter suggested. "Isn't that what advertising is for? To make people want something they never knew they needed? Maybe there's this great, untapped need in the world to drink tepid milk vomit, and you're the woman to fill it."

"Thanks," I muttered, then fed my failure to the dogs. Kowhai whimpered and refused to eat it, and this is a dog who considers roadkill a delicacy, second only to the taste of her own poop. I felt a little offended.

I went back to the books, and on the third attempt, I solved the mystery. The American recipe I'd been using called for a specialty cheese culture with the rennet already included. Since that fancy product isn't available in New Zealand, I'd have to add my own rennet. Three drops of dead cow juice made all the difference, and the next day my milk had firmed into a solid white mass surrounded by a clear and cloudy whey.

"Wow!" Peter marveled, observing my squishy white lumps. "You made cheese!"

"Not yet," I corrected, scrutinizing my work in progress. "First I have to age it."

Once I started aging it, my goat cheese grew lots of mold. There was a bright yellow mold that looked as though a child had scribbled on the cheese with a highlighter, and a hot pink mold in a poisonous, festive hue. Then there were the gray fuzzies. I consulted Ricki Carroll's classic *Home Cheese Making*, and she had plenty to say on the subject. "That mold is called '*poil de chat*,'" she warned, "and you do *not* want it on your cheese."

"Crap on a stick," I muttered one morning, rummaging in a drawer for a rag.

"What's wrong with you?" Peter wanted to know.

"I have *poil de chat*."

"You have what?"

"Cat hair." I thought about this for a minute. "Although *chatte* is a rude word for vagina, so I guess you could call it pubic hair."

"Your *cheese* has *pubic hair*?"

"Please." I held up my hand. "It's *poil de chat*."

Everything seemed to be falling apart. My cheeses were rotting away in a pile of multihued mush. Silas couldn't be left alone for an instant, because any moment he could have a seizure. Each day, I expected to get a note in the mail politely but firmly informing us that Purua School could no longer meet Silas's medical needs.

Then, one evening, the phone rang. "Antonia?" Sophia's voice came through the receiver. "We need to talk about Silas."

My heart sank. This was it, the call I'd been dreading. "It's just that Patrice only stays with Silas until two-thirty each day. So there's that last half hour of school where no one's watching him. And if he had a seizure, and hit his head . . ."

Her voice trailed off. I could see where she was going with this. "You need me to start collecting him early, is that it?"

"Yes, I'm afraid so. I just really can't have him out of sight for right now. While we're getting the fits under control."

So he wasn't kicked out. Not yet. But we both knew there was only one place where Silas would be sure of constant supervision. Short of a hospital, it was the special-needs school.

I heaved a deep sigh. Both our chances at a house had fallen through, we had only two more months on our rental, and after that we didn't have any idea where we'd go. And we still had nineteen animals to care for.

I was reminded on a daily basis precisely how many animals we owned—not just because Rebecca and I were the ones to feed and water them, but because they had now begun to roam free. The fencing on our rental property was breaking down, and the animals just strolled through the gaps. One day, I'd see a calf contentedly munching leaves off a passion fruit vine, and the next I'd see the chickens jabbering and scratching under the trampoline. Becca and I chased them back into their enclosures, using treats and strong language, and they'd escape again. Peter tried to patch the fencing with odd bits of lumber, but it needed a complete overhaul, and that wasn't something we could do for a rental.

The calves and the hens weren't the only ones who went rogue that spring. Moxie and Stripe, nimble since the day they were born, started leaping on top of our cars. Their jumping was better than their balance: when they landed on a roof, they would skid, hurtling down the windshield and landing in a heap on the car's dented hood. The goats thought this was hilarious, and they jumped on the cars all day, dislodging our license plates and shredding our windshield wipers.

Occasionally the mayhem outside overwhelmed me, and I'd retreat to my bedroom while Becca lay down on her nail bed. Rebecca could lock the door to her sleep-out, but for me it was rarely restful. Either Silas or Miranda would jump on the bed, or a goat would follow me inside and head-butt me gently as I was drifting to sleep. To top it all off, the dog still had her period, and soon the carpet was speckled with droplets of blood, as though we were living with an organic roommate who walked around in the nude and eschewed tampons.

And then Ba fell out of favor, at least as far as most of us were concerned. Peter and I saw him as an overgrown sheep with crusty horn buds and giant testicles, but Rebecca still considered him her baby. Ba must have weighed eighty pounds by now, so it began to look ridiculous when Rebecca sat down and coaxed him onto her lap. Mostly she did this outside in the grass, but one unusually chilly day in November, she invited him inside the living room. Peter was working late, so we'd fed the children early, and I was standing at the sink, getting started on the dishes. Rebecca crooned a lullaby in the giant sheep's ear and lay back on the carpet, allowing Ba to splay across her chest like an obscene, furry baby. Legs spread to make room for his back hooves, she shut her eyes and dozed, stroking his thick white fleece.

This must have been very relaxing for Ba, because he then produced a large, shiny bowel movement on the living room rug.

These were not the darling little lamb berries he'd scattered in his youth, but a generous, rippling sheep turd sitting moistly on the carpet.

Busy with the dishes, I didn't notice a thing until Peter came home. He slammed down his laptop case. "That's *it*," he roared. "Out! Get that thing out of the house! No more sheep inside! And clean up that mess!"

Rebecca complied, but for the next few weeks, Ba remained in disgrace. We decided not to bring him to Calf Club that year. Silas wouldn't be able to compete with him, not now when his mind seemed so tenuous, his language all but gone. So we left Ba in his pen and went out to support the other children.

Calf Club Day fell on a foggy spring morning in November, and all of Purua turned out for the occasion. Karl and Catherine sold sausages and raffle tickets for the school, Patrice made hot coffee and sold cookies from the staff room. Autumn was selling trees for people to plant on the school grounds, and jars of yummy, scrummy honey that the children had harvested from their hives. Nick and Amanda were there, Nick with his digital camera, getting ready to film the animal judging.

Sophia darted around in the role of hostess, making sure the parents had comfortable places to sit and that the children tied up their animals properly. She wore a green paisley tunic in a gossamer Indian silk, and it swirled around her as she walked. "Oh, you're *wonderful* to come after everything that's been happening!" she squeezed my hand when she saw us. "And how are *you*, Silas?" she asked, bending down to his level.

"Bus," he replied.

"Yes, you do love the bus, don't you, dear?" She patted his head affectionately. "And did you raise an animal this year, Silas?"

"Bah . . . mmmm," Silas said.

"That's right, kid. Our lamb's name is Ba." To Sophia, I explained. "We thought it was a name he'd be able to pronounce."

"Yes, yes, of course," she murmured distractedly. "Now I must go see about the cake auction . . ." She whirled away just as Rebecca approached, Miranda in tow.

"Mama!" Miranda announced. "I'm hungry! Can I have a sausage?"

"Sure," I nodded, heading over to the barbecue. I bought sausages slathered in sugary tomato sauce for both the kids. I got one for myself, too, with ketchup and a bright yellow mustard I hadn't seen since childhood.

"Want a fizzy drink?" Karl grunted, and I fished out a cold Coca-Cola from the cooler.

"Sorry, they didn't have anything vegetarian," I told Rebecca, handing out sausages to the kids.

"That's all right." She shook her head. "Patrice gave me a piece of cake when we got here."

"All the lambs are tied up on the fence!" Miranda announced, finishing her sausage. "Come and see, Mama! It's really cool!"

I went back with them to check it out. All the young animals were tethered at the far side of the playing field: eight or ten sheep, a large doe-eyed Jersey calf, and a small gray goat named Cole. Rebecca was inspecting the competition, shaking her head in disappointment.

"Ba would have *wasted* these lambs," she whispered. "He would have won, hands-down. I can't believe we didn't bring him."

"I'm sure you're right," I told her soothingly.

The judging began, each child asked to line up with the animal they'd raised themselves. Maria's husband, John, was the chief judge. Wearing a stiff button-down shirt and holding a clipboard, he appraised all the animals with professional detachment. John was one of the real farmers, and he wasn't about to cut these kids any slack.

We stood in the audience, watching the judging. The younger students were leading their animals through an obstacle course, and Amanda's daughter Amelia seemed to be struggling. Her lamb sat down halfway through, his head hanging stubbornly low. Amelia pulled with all her might, but Blake was bigger than she was. He wasn't about to budge. Tensing his jaw, John made a note on his clipboard.

"G'day." Lish sidled up to me, Skin beside her. He nodded his head and winked, then raised his hand to Peter.

"Hey, you guys." I grinned. "Thought you'd be out there with the animals, Skin. Looks like Amelia could use a little help."

He held up his other hand then, wincing. "Put a fencing wire through my thumb," he told us. "Bit sore an' that." He grinned, his brown eyes shining in that weathered face.

Peter looked concerned. "Did you go to the hospital?"

Lish rolled her eyes and grimaced, but Skin just shook his head. "Went to Emergency, but they took too long. Bloody doctors. Buggered off home, gave it a yank with a pair of pliers."

"But what if it gets infected?"

Lish threw up her hands. "I keep trying to tell him. But will he listen to me? No, he will not."

"Nah." Skin batted away Miranda, who'd started charging him in the thighs. "She'll be right. Just pour on a bit of meths, good as new." He nodded at me.

"What's meths?" Peter whispered.

"Paint thinner," I said.

Lish touched my arm. "Heard yous had a problem with Silas. I'm off in the afternoons. Only work till two. I been looking after him, that last half an hour."

"But"—I stammered—"Really? I can't . . . I mean, we can't afford to pay you."

Silas pulled on my hand. "*Toit*," he announced.

Lish waved me away. "Don't need pay. We have fun together! I throw him a ball, make him run for it. Tire him out. Eh, Silas?" She gave him a high five and nudged me in the ribs. "That's what mums need."

"Toit!" Silas repeated. "Toit! *Toit!*"

"Mama," Miranda said softly. "I think Silas needs to go to the toilet."

"We're going," I told Silas. Then I turned back to Lish. "Well," I said cautiously, "I mean—thank you. He'll be able to stay in school the whole day, then."

Lish just grinned. "No worries. That's how we do things here. Now get that boy to the loo!"

I guided Silas into the staff toilet, because it was a wide, comfortable room where I could easily help him with his pants. I sat him down on the toilet, and he shrieked, even though he had to go. Then he fought me when I tried to help him back up again. Turning on the water so he could wash his hands, I caught a glimpse of myself in the mirror. Bright yellow mustard and brown Coca-Cola stains spattered the front of my shirt. I looked like a drunk five-year-old. And I still had the ridiculous pair of bunny ears on my head, insisting to the world that life was a fucking breeze.

When we came out, Patrice was standing there in the corridor examining an assortment of school pictures from the past few decades. The photos were all black and white, the kids' hair combed back carefully, rigid smiles pasted on their bright little faces.

"Who is that?" he asked playfully, pointing at one little boy.

I looked more closely. About ten years old, the boy had thick hair that stood stubbornly on end and a wide, cheeky grin on his face—not the dutiful smile of his classmates.

"Oh my goodness." I caught my breath. "It's Hamish." He was just a goofy little boy, with the same easy smile I'd seen the night I caught him drinking a beer. I reached up to adjust my bunny ears. "I wonder what happened? To make him so serious?"

Patrice shrugged. "He just grows up," he remarked with a tilt of his head. "As we must do."

At the end of the day there was a cake auction, a spectacular display of high-calorie skills. Arrayed on the wooden lunch table were rainbow cakes, chocolate cakes, and coconut creams, but the one that caught my eye from the start was a homemade apple pie.

"Thirty dollars," I bid, raising my hand in the air.

The auctioneer was a local farmer, in gumboots and a dark leather hat. "Thirty, thirty, do I hear thirty-five, thirty-five, thirty-five, do I hear forty?"

"What?" I looked around accusingly. "Who bid on my pie?"

"We're American," Peter argued. "We should totally get this."

"Forty dollars, going once, going twice—"

"Fifty dollars!" I shot my hand in the air.

The auctioneer looked amused. "Fifty dollars for a pie. This must be the best apple pie in the world."

"Fifty-five," said a voice behind me. I turned around to see Skin with his good hand in the air, a saucy grin on his face. "My pie," he mouthed, jabbing at his chest with one thumb.

Beside him, Lish had her head in her hands. "I could buy you *four* pies for that," she mourned through clenched teeth.

THE GRAB-AND-YANK

"Sixty dollars!" I yelled, holding my hand up.

"Sixty dollars, do I hear sixty-five? That's sixty dollars, going once, going twice, *sold* to the lady in the bunny ears."

"You got my pie!" Skin hissed, pretending to be angry. "I want my pie!"

And strangely, that was the last thing he said to me.

CHAPTER FOURTEEN

OUTNUMBERED

It was a good thing Ba stayed home that day, because Calf Club 2013 was a death trap for sheep. Within three days, Amelia's lamb Blake was dead.

"Bloody rhododendrons," Amanda told me over the phone. "We tied him up next to the rhododendron bush at Calf Club, and the next thing we knew he had this lime green froth coming out at both ends. Then it was coming out his nose, too, like a bright green mucus. We took him to Jackie, and she gave him an injection. Amelia read him a story. But there wasn't anything we could do."

"What's a rhododendron?" I knew it was a plant thing, but I couldn't picture it in my head.

"Those bright pink flowers, the ones growing at the end of the playing field."

"But there were all those farmers there! Didn't anyone tell you the flowers were poisonous?"

"It's not like that, in the country. Animals just die. You just have to get used to it, living out on a farm."

I understood her point, but it still seemed like the death plants might have been pruned before little kids started tying up their lambs nearby.

I didn't have much time to worry about dead lambs, in any case, because now I was working on kimchi. Brewing a jar of fermented cabbage seemed like a perfectly reasonable thing to do, after making wine and cheese. All these concoctions are full of germs and bugs that break down the food for you, making something ordinary into a delicious dish that's pungent and strong. Working on the back deck with my bacteria and molds, I liked to imagine millions of little sex slaves busily reproducing and turning fruit juice into wine.

The basic Korean kimchi ingredients are cabbage, garlic, ginger, and scallions, and from there you can just go crazy. The recipe I tried called for hot chili peppers and dried shrimp, which I managed to locate on a dusty shelf at the back of Whangarei's lone Asian grocery. The shrimp looked like dead beetles curled up in little knots. I opened the bag and sniffed. An acrid, deep-sea scent so strong it seemed to have dimension and mass slapped me square across the face. I snapped the bag shut.

The shopkeeper chuckled. "Very strong," he reminded me. "Just use very little bit."

I paid for my shrimp and got back in the car. Miranda, who loves doing the weekly shopping with me, was instantly transfixed. "Mama?" she asked. "What's that?"

"Dried shrimp."

"Can I try some? I want to try some. Please? Mama, can I please try some?"

So I gave her a shrimp.

"Mmmmm," she said delightedly, placing the shrimp on her tongue. Just as fast, she spat it out again. "That's disgusting, Mama. That's not a shrimp."

"I promise you," I reassured her, "it is. And we're going to take it home and make kimchi."

When we got home, Rebecca was resting in her sleep-out. She seemed to be spending more time there lately. With just six weeks left to go in her visit with us, it felt as if she were starting to separate from our family. That, and her nail bed was more relaxing than the chaos outside.

I started the kimchi by brining some cabbage, soaking it in a strong saltwater bath and weighing it down with a heavy milk jug. Silas staggered out of his bedroom holding a plastic elephant in one hand. He stumbled awkwardly, as though the room were tipping around him. The seizure medications he was on seemed to mess with his sense of balance. He held up the elephant.

"Eh-faaaah . . ." he told me, with a shy smile. "Eh-faaaaah . . ."

Peter, who was sitting by the window watching the baby goats jump on my car, glanced up. "He used to say that word better," he noted. "It used to sound more like *elephant*."

"I know." I sounded calm, but that's because I was carefully tamping down the crazy lady inside me, the one who was burning cars and screaming. I poured a glass of peach wine and drank. "I don't know what to do about it."

Silas perched his elephant on the top of a wooden dresser, then tilted his head to view it from a new angle. Peter turned his attention back to the goats. "They're going to break your windshield eventually," he pointed out. "You realize that, don't you?"

"But they're *so cute*," I protested, which seemed like a watertight argument. Moxie took a flying leap at the rear of Peter's car, skidding across the roof and hooking the radio aerial in one hoof. She slid onto the hood and tore the aerial out of its base. Then Stripe leaped up and started chewing on it.

"That's it. Those goats are going down." Peter pulled open the sliding glass door and stepped out on the deck, Silas right behind him.

"Eh-faaaaah . . ." Silas chanted.

"That's right, Elephant Man." Peter took Silas's hand. "Let's go."

Deciding it was time for a break, I took my wineglass out to the deck. Rebecca had emerged from her sleep-out and was sitting there on the steps, cuddling Ba. Little piles of sheep shit dotted the planks.

"That's so sweet," I commented, and I meant it. He was an overgrown animal, but Ba still nestled his head into Rebecca's neck, just as he'd done when he was a baby.

Then he stiffened and vomited a stream of rotten grass slime down the front of her shirt.

"Ugh!" She jumped to her feet, grimacing and holding the stinking wet shirt away from her.

"Meeeh," Ba bleated, annoyed at being dumped on the ground.

"Becca?" Miranda came out on the deck. "My body's talking to you, and she wants you to give her a cuddle."

"What?" Turned away from us both, Rebecca was frantically rubbing the front of her shirt with a rag. "I think I have to change. This stuff is really stinky."

"Ew, Mama," Miranda sniffed the air theatrically. "It smells like poop."

Giving Rebecca some privacy, I took Miranda's hand and walked her inside. "Not poop, Magnolia. Just grass vomit. Let's try to be accurate."

The cabbage was still brining, so I turned my attention to the ginger bug, which is the starter for ginger beer, much like sourdough starter is the base for the bread. The finished drink is not alcoholic, but the idea is to ferment it just long enough to form natural carbonation. I started grating my ginger, mixing it with the leftover whey from Pearl's cheese and adding sugar to feed the live cultures.

"What are you making?" Miranda wanted to know. She was playing with an empty soapbox on the floor.

"Ginger bug," I told her. "We'll use it to make a fizzy drink. It's delicious."

"Ginger bug?" Miranda considered this. "You can have my bug if you want, Mama. I think it's a little bit dead, but you can have it."

She handed me the soapbox, which appeared to contain a single dead fly. "Oh, thank you, Magnolia," I told her, placing the box on a high shelf. "That's just what I needed."

Peter walked back in the room then, wiping his hands on his jeans. "I tethered the babies," he announced. "No more dancing goats."

"Where's Silas?"

Peter turned and looked behind him. "I don't know. He was out with the goats, last I saw."

A piercing shriek rang out from the paddock. Silas came bolting on the deck, Ba close behind him. *"Home!"* he shrieked. "No. No. No! *Away!"*

I looked over at Peter. "What's he so upset about? It's just a sheep."

Ba answered my question by slamming his head against Silas and shoving him against the sliding glass door. I threw down my grater. "Save him!" I screamed.

But Peter was already there. He flung open the door, grabbed hold of Silas, and kicked Ba out of the way.

Ba fell back on the deck, bleating irritably.

"Silas!" I grabbed my son. "Are you okay?"

"No, no," he sobbed, red-faced and frightened. "Away!"

I inspected him closely, but he didn't seem hurt.

"Mama?" Miranda tugged at the back of my shirt. "Ba did bite Silas?"

"I don't think so. I think he's just scared."

"And Ba would eat him all up," she decided. "And you will never see him again."

"I don't think it's quite that bad."

Peter watched Ba retreat off the deck. "Why's he so aggressive?"

"Well, he did just throw up on Rebecca. Maybe he's not feeling well." I was still rubbing Silas's back, as his sobs turned to shuddering breaths. "Come on, Miranda," I took my daughter's hand. "Let's go collect the eggs."

This was usually a fun part of our day, when the three of us went down to look for eggs in the chicken nests. Gathering eggs is one of the harmless, picturesque parts of farming, the sort of gentle activity we'd imagined when we first moved to the countryside.

Except now it was a blood sport. Rogue animals roamed the property, lusting for fresh meat. At the very least, they threatened to jump on our cars, frighten the children, or charge us head-on. We tiptoed down to the chicken coop, dodging Pearl and her babies on the way. Cinnamon and Lil' Lady were under the quince tree, munching grass and watching us warily.

Miranda stepped into the chicken coop. "I think there will be five eggs today, Mama. What do you guess?"

"Wait, Magnolia," I urged. "It's best to take the garbage lid with you, just in case—"

There was a flapping sound. I looked up to see Jabberwocky pouncing, talons extended, looking to seize my daughter in a cloacal kiss.

"Ow, Mama!" Miranda screamed. "Help!"

"That's it, you bastard!" On instinct, I grabbed a discarded two-by-four that was leaning against the side of the coop and lunged inside, beating at the bird with one end. "Miranda," I yelled, "get outside. *Now!*"

Still screaming, she scuttled out of the coop, and I slammed the door behind her.

"Mama?" Miranda asked. "Next time? If Jabberwocky does get me? Then will you get a gun and dead him?"

"It's looking that way," I told her. "It's just a matter of time."

We had dinner at Autumn and Patrice's that night. It was a simple meal, with corn from their garden and mussels that Skin and Patrice had picked off the rocks at the seaside the day before. Patrice was lit up and grinning, tending the wood fire to grill our mussels. "It was so beautiful. It was his land—tribe land. You cannot go there unless you are with someone from that Māori tribe. And there was no one there!" He prodded the embers and blew, nursing them back to life. "The water was so clean you can't believe it."

"So Skin's Māori?" I hadn't really thought about this before. Tribal identity isn't race based in New Zealand—some Māori are blond and blue-eyed, and a person with dark skin might be from anywhere: Papua New Guinea, Fiji, or even South America.

"Yep, Ngāpuhi* I think. They've invited us back tomorrow," Autumn said, her pale skin milky in the fading light. "There's no

---

* Ngāpuhi (NA-pooey) are the largest Māori tribe in New Zealand, their territory extending across the northern part of the North Island.

school on Monday, so we're all going to camp. Skin says he'll show Titou how to catch a big snapper."

It sounded great. I'd so much rather think about catching fish and grilling mussels than the mess of wild animals I had back home, or the storm systems raging in my son's brain. I looked over at Silas, wrapped in a blanket on the grass. He always seemed so tired these days.

"I think we have to kill our sheep," Peter said, changing the subject. "He went after Silas today."

"And the rooster," I reminded him. "Actually, all the animals are getting a little scary. I'm starting to feel outnumbered."

Autumn sipped her wine. "You know, they're not meant to be roaming free. We do have fences here in New Zealand."

"I know, I know." I sighed. "But the fences keep breaking, and then they escape."

The embers were now white-hot, and Patrice started placing mussels on the grill. They popped and sizzled, releasing a smell of wood fire and briny ocean. "That rooster is a motherfucker," he observed in his thick French accent. "A rooster like that, coming to a small child—it's like fighting a super villain with claws and a beak. You remember, Autumn?" he asked. "When your mother pulled the head off a rooster?"

Peter looked impressed. "Straight up ripped its head off? What did it do?"

"Oh, God." Autumn laughed. "She didn't mean to. She was trying to wring its neck, and she wanted to do it right on the first go. So she pulled a little hard, and the head came right off in her hands. She was just standing there, holding a bloody rooster head. It was awful."

Peter winced. "I think we'll use an axe on Jabberwocky."

But Autumn wasn't finished. "You think you lot have trouble with chickens. The next chicken she had was standing in the doorway of the coop, and the door blew shut and squashed its head in. Then her dog got the last one, and she thought, *Shit, I don't have any chickens.*"

The mussels had all popped open. These were meaty, green-lipped mussels, each one the size of a small steak. Patrice set them on a platter with tongs, the tender flesh bubbling in the simmering juice.

Autumn went on with her tale. "So she rang up the free-range chicken farm to get some more chicks, and the people asked her what had happened to the chickens she had, and I guess she was too honest, 'cause she told them. And they said, 'Ah nah. We've got no chickens for the likes of you.' She couldn't even give them money!" Autumn reached for a mussel. "It was so embarrassing."

I picked up a scorching shell and slurped down the meat inside, an exquisite combination of hot smoke and sea.

"'No chickens for the likes of you,'" I repeated. "Somebody should have said that to us years ago."

The next morning, I got back to work on my cheeses. I had a few jugs of Pearl's milk on hand and I wanted to try making a Camembert with goat milk, rather than the more traditional cow's milk variety. But my cheese boards, squares of kauri that Peter had carefully cut and sanded for the purpose, appeared to be glued together. I peered more closely. They were cemented with clumps of pale brown mortar, and when I twisted the boards apart, the "mortar" crumbled. Dozens of dead spiders tumbled out, among them, several glistening gray larvae.

I ran for the phone. "Autumn! What the . . . what the wha . . . ?" I stammered, traumatized at the Halloween horror show taking place on my back deck.

Autumn just laughed. "Ah, those are mason bees. They're everywhere. Any little crack or crevice you forget to clean, they'll make their nest." This sounded familiar. Sophia had told me about mason bees, but at the time, I hadn't give it much thought. Now they were horribly real.

"But why is the nest full of spiders?"

"They lay their eggs in the spiders. Then the larvae eat the spiders from the inside out. It's all completely natural."

Maybe to some people. I changed the subject. "Hey, can you come over this weekend to take a look at Pearl's tits? I think she might have mastitis."

"Ah, I'd love to, but I can't. We're going out to Skin's, remember?"

"All right." I sighed, resigned to a solitary weekend of sick sheep and dead spiders. "Maybe we'll see you when you get back?"

"Ah, definitely." We said our good-byes, and I hung up the phone.

Pulling on a pair of gloves, I retrieved the contaminated cheese boards and brushed off the doomed spiders, then submerged the whole mess in a sinkful of boiling water. I poured a cup of coffee and went out to the deck.

It was a quiet weekend. We moved through the rhythms of farm life, collecting eggs from the chickens, milking Pearl in the morning and checking her milk for mastitis. Saturday afternoon, Peter gave Silas and Miranda a ride on the lawn mower while I cleaned sheep shit off the front deck. I washed my hands and flipped my goat milk Camembert, thinking contentedly about the microscopic cultures I was colonizing on their surfaces.

Sunday was blistering hot. Rebecca took our car and drove out to the beach. Peter and I sat on the back deck in the evening, sipping peach wine and watching the wood pigeons streak through the totara tree. We put on a DVD for Silas and Miranda and slipped away, walking down by the creek with our glasses of peach wine. It was cool down there, in the shade of the great trees, the glossy green shrubbery and tall orange wildflowers crunching underfoot. We undressed at our place beneath the tree and made love in the karamu like furtive teenagers. Just above our heads, not ten feet away, the chickens clucked disapprovingly.

Peter turned on one side, propping his head in his hand. He nuzzled my neck and nibbled my ear, his beard tickling my cheek. "Can we still come here?" he whispered, "when Katya and Derek come back? Do you think that would be weird, if we just walked on their land and headed down to our place in the grass?"

"Yes," I concluded. "Although, if we don't find somewhere else to live, we might be pitching a tent down here."

We pulled on our pants, our skin still damp with sweat. I reached over and brushed a tuft of grass from Peter's eyebrow. Our wineglasses lay forgotten in the leaves.

CHAPTER FIFTEEN

CATCH THE BIG ONE

"How is it?" Rebecca asked as I eagerly opened my first jar of kimchi. This concoction was made with heirloom cabbages and daikon radish, scallions and garlic from the farmer's market, crushed ginger and expensive Japanese chilies, with kelp and black sesame seeds, not to mention the tiny dried shrimp—so I was expecting more than just cabbage. I was expecting epiphany.

I cracked the lid and sampled a bite. "Salt," I gagged.

"It needs salt?"

I bent over the sink and spat out the kimchi. "No, it tastes like salt. All those exotic ingredients, and I made a jar of salt."

"Are you sure?" Becca lifted the jar and inspected it. "Maybe that's what it's supposed to taste like."

"Not unless it's a brand-new way for Kim Jong-un to torture people. This stuff'll burn a hole in your tongue." I dumped it in the compost bin, certain even the chickens wouldn't eat it. Then I poured myself a cold glass of ginger beer and sat out on the deck to recover.

The ginger drink slid down my throat, sweet and sparkling, with a gentle burst of spice. Teeming with microorganisms, it felt alive. Drinking it made me feel virtuous somehow, as if I'd just gobbled up a bowl of flaxseeds and nonfat, organic yoghurt. I took a deep breath, beginning to relax. And then I noticed that Silas was playing with shit.

For once, it wasn't his. Happily perched on the deck's front steps, he was picking up goat berries and flinging them at a passing chicken. Admittedly, the berries were dried, so they did look like dark brown marbles—but still. I remembered the incident with the vampire worms and shuddered. "Silas," I started to correct him. "I don't think you should—"

"*Ew*, Mama, I stepped in poop!" Miranda howled, and I turned to see her pointing the sole of her foot at me, a generous helping of chicken poo dripping from between her toes.

"Hang on," I muttered, going to the kitchen for a rag. "And stop throwing goat turds!" I called over my shoulder to Silas.

Inside, the phone was ringing.

"Hello?" I answered.

"Antonia, it's Autumn." My friend's breathless voice came through the phone. It seemed odd for her to be calling now. Weren't they out at the beach camping?

"Is everything okay? What happened?"

"Skin has died."

"*What?*" At first I didn't process what she'd said. It wasn't possible. "Are your children okay?" I blurted out.

"I think so. Yes. I don't know. I mean, I think so." Autumn was breathing hard, this woman who was always so calm and unflappable. There was background noise on the line, street sounds. She sounded as if she was running.

"What happened?"

"I don't know. Patrice just called. He might have told me what happened, I don't know. When he said Skin died, I didn't really hear anything after that."

"*Mama!*" Miranda howled from the deck. "I still have poop on my foot!"

"Were they hunting? Was there a gun?"

"No. But I think they were night fishing. Skin might have gone out alone."

"No, Silas, *no!*" Miranda sounded panicked. "Stop throwing poop on me!"

Autumn's voice came back on the line. "Look, I'm walking to my car now. I have to go get them. Can you pick up Titou for me from school?"

"Of course, no problem. Should I tell him what happened?"

"*No,*" she practically yelled into the phone. "Don't tell him. I'll tell him myself."

"Can I do anything else?"

"No, I don't know. I just have to get out there." With a click, the phone went dead.

"What happened?" Becca asked when I'd hung up.

"*Mama!*" Miranda was screeching now. I told Rebecca the news as I went for a rag, then hurried outside to rescue my daughter. I pulled Miranda on my lap, holding her close as I wiped down her foot. "Magnolia," I said into her ear, "I need to tell you something." I braced myself for tears. "Skin has died."

"Yeah." Miranda nodded as though she'd already heard the news. "And he is never coming back."

"No. He isn't coming back."

"And, Mama?" I waited, thinking maybe she'd have an important thought about mortality. "There is really poop all over the deck. You should clean it."

I looked around. Somehow, without realizing it, things had gotten out of control on our little farm. The front deck, with its long outdoor table where we gathered for barbecues, was strewn with barnyard waste. There were land mines of wrinkled sheep shit, a smattering of goat berries rolling around on the floorboards. Among these obvious threats, the occasional sneak attack lay in wait: a chicken poop, watery and pale, its light brown color camouflaged in the wood. Beyond the deck, the driveway was littered with dog droppings, and I could just make out one of our calves defecating beneath a peach tree. I heard a scattering noise, like buckshot, and I looked up to find that Silas was still tossing goat berries.

"Silas, *please*, I—"

Hamish's quad appeared at the top of the drive. "Oh, *great*." I snatched the poop out of Silas's hand. "Hamish is gonna *love* this."

But Hamish didn't mention the manure on our driveway, the cow in the peach trees, or the sheep on our deck. He glanced down at Silas, but he didn't get off his quad.

I stepped forward to greet him, and what I saw gave me a shock. Hamish was in his same green coveralls, his strong hands smudged with farm dirt and clutching the wheel. But his eyes were red and full, tears about to spill down his cheeks.

"Sad news about Skin," he managed. His voice caught, and he coughed. "The whole district's in mourning."

"I just heard," I said. "Do you know what happened?"

"I know he was out there with his mates." Hamish smiled. "Catching fish, out on the beach. Would have been the best way to go, for him. Out on the beach in the sunshine. Still"—he coughed again, blinking as if something were caught in his eyes—"he was bloody young."

"Fifty-two."

"Yeah." Hamish started up his quad again. "Makes you think."

"Mama?" Miranda tugged on my shirt. "I'm so sad that Skin died."

I picked her up. "I am, too, Magnolia," I whispered in her ear. "I am, too."

I felt a tug on my shirt and looked down. There was Silas, with his last ball of goat poop. He held it up to me as though it were a treasured gift. "Home," he said. "Poo."

I took the goat berry and flicked it away. "Yep, Silas." I nodded. "That's about right."

I called Peter at work to tell him what had happened, and I spent the rest of the day worrying about our friends. Whatever had happened to Skin, it must have been violent, and Nova and Maris were there. This wasn't the fun blood, the kind you can wash out with hot water and soap. They were stuck at the beach with a body. And they were just little girls.

I cleaned off the deck in the afternoon, watching through the sliding glass doors as Titou and Miranda played together inside. They were pretending to be rocket ships.

"First I get my fire on!" Titou hollered, flexing his arm muscles and bending at the knee. "And then one . . . two . . . three . . . *go!*" The two of them tore through the house, running at top speed down the corridor and then leaping onto my bed. Through the window, I could see them jumping up and down. Silas sat on the couch watching, his Dart pressed to his ear. He didn't join in, but he had a huge grin on his face, giggling every time the other two screeched by.

I swept the poop off first with a push broom, rolling the goat berries out in front of me until they tumbled to the grass, scrubbing harder at the sheep and chicken poop, which seemed to retain the adhesive qualities of stinking brown cement. The problem, I reflected, was that the fencing was so terrible here. Katya and Derek didn't keep many animals, so they'd allowed their fences to break down over the years. Our sheep and goats just strolled on through, which was why they were always on the deck, crapping and waiting for a cuddle, wanting to be near us. They were all herd animals. They didn't want to be left alone.

I filled a bucket with hot water, swirling in a generous slosh of bleach. Everything out here grew with abandon: the livestock, the fruit trees, the grasses, the worms and molds and bugs. Everyone needed his own safe place, even the animals. They had to be controlled and contained, or they'd trot up your front steps and shit on your deck. I took a stiff brush and scrubbed at a chicken poop. Inside, Miranda and Titou were now standing on the couch tossing Legos at Silas, who was crawling around on the ground.

"Bad doggy!" Miranda shouted, hurling a handful of Legos. "No biting!"

I supposed I should have been angry that they were pelting my disabled son with toys, but actually I was pleased. If Silas was crawling around on the ground, he was halfway to pretending to be a dog—and that meant he could imagine. Which was good news for his brain.

Autumn arrived just as I finished scrubbing, and Titou ran out to meet her, flexing his muscles. "Mama!" he cried. "We are going back to the beach now!"

"No, Titou." Autumn's face was flushed, and her cheeks were blotchy from crying. "We aren't going there. Skin has died."

When he saw his mother's tears, Titou's little face crumpled. "No!" he protested. "I have to catch a snapper! Skin will show me!"

"No," Autumn said gently. "Skin is dead now."

She scooped him up in her arms and sat with him on the deck, nuzzling his hair with her face. Then she told me what had happened.

"We had such a beautiful day yesterday. Fishing and playing on the beach. We had a big fire at night and cooked the snapper we caught. Then I drove back to town with Titou because I had to work today. Patrice stayed out there with Lish and Skin. And the girls."

Titou wriggled in her arms, and she let him go. He ran inside to find Miranda.

"Then this morning, they were in the shed, picking out their fishing poles, getting their bait ready. Skin sat down to drink a coffee with Patrice, and he just went into a seizure." She smiled faintly. "Of course Patrice knew what was happening, because he'd seen it before with Silas.

"Skin fell right over. Patrice turned him on his side, but he was gone. They gave him CPR. Patrice gave him CPR for half an hour or something. Then Lish took over. They tried to bring him back for more than an hour. He was gone. It was just a massive stroke."

"And the girls were there?"

"Yeah. They haven't talked about it much, but I think they saw everything." The tears spilled over her cheeks as she talked. "They had no cell phone reception there, so finally Patrice just ran. He ran to a farmhouse and rang me, and I—I didn't know what I was doing. I think I blacked out."

"You must have been scared for your kids."

Autumn shook her head. "Maybe, I don't know. I don't remember it, to be honest. I just had to get to the beach. And when I got there"—she took a breath then, a deep, shuddering sigh—"he was gone. He was just laid out on the floor of the shed, and Lish was lying there with him, cuddled up to him, just weeping. There was so much pain."

I thought about this for a minute. "Remember that wire he drove through his hand?"

Autumn nodded, smiling.

"He walked out of the hospital before they could see him. You know they would have taken his vitals. If he'd had high blood pressure or something, they might have caught it."

She shook her head and laughed grimly. "I thought of that. And I think he knew. His dad died early, of a heart attack. I think he knew they'd find something, and he didn't want to know about it. He just wanted to go fishing."

Three days later, we went to the wake. It was hard to think about anything else. I spent the intervening hours starting another kimchi, turning my cheeses, decanting my raspberry mead. The problem with the first batch of kimchi, I conceded, was that I'd made it while drinking my powerful peach wine, so I read the recipe wrong and dumped in too much salt. The trick to most ferments is to make a safe place where the culture can flourish. You don't want a sterile environment, like an operating room, because nothing can live there. Instead you adjust the heat and the salt and the acids so that one group will thrive and push out the chaos of all the rest—the sheep shit, the pubic hair mold, the worms and the maggots and the death.

It was comforting to do my farm chores each day. When Peter and the children left for work and school, Rebecca spent time in the paddocks while I took kitchen scraps to the chickens, dodging Jabberwocky as I scattered their food. In the aftermath of numerous sexual assaults, one of the hens had gone broody, and she now sat jealously on a clutch of warm eggs, squawking a warning if anyone approached. I milked Pearl, checking her milk for signs of mastitis, and gave a quick pet to Ba and the cows. The alpacas glared at me, but at least they never charged.

And I turned my cheeses. Four moist, white bricks of goat milk sat ripening in my cheese fridge, and each day, I turned them, inspecting them closely for multicolored mold. When I saw some, I rubbed a little salt on the lesion, carefully cleaning the draining racks and the ripening box with hot soap and water.

The day of Skin's wake, I decanted the raspberry mead, noting with satisfaction that the wine yeast, *S. cerevisiae*, had reproduced splendidly, converting honey, water, and fruit into formidable booze. I sipped a little off the rubber tube to create a siphon, and with just one gulp, I wasn't sure if I should drive.

So I gave the keys to Peter, and he took the wheel. We packed the children in the car, and Rebecca squeezed in between them. I had a pot of hot chicken soup and a couple of bottles of homemade mead to share.

"I just thought we'd know him forever," I kept repeating as we drove. The sky was overcast, the day heavy and humid. "I liked him so much, and I knew we wanted to stay, and I thought we'd be friends for ten or twenty years. I thought we had so much time."

Peter nodded, saying nothing.

The wake was at Skin's mother's house, in a tiny seaside community called Whananaki. We drove a winding road to get out there, giant ferns and native trees pressing in on both sides. When we emerged from the bush the road straightened, and we pulled into a coastal village. Cars were parked everywhere. There wasn't any need for an address or even for directions—Doreen's house was overflowing with people.

It had started to rain, and we got out awkwardly, pulling on raincoats and unfolding umbrellas. Most of the faces we saw were unfamiliar, but everyone smiled grimly and nodded.

Inside, there must have been two hundred people. Lish was there, looking pale and shocked. I gave her a fierce hug, then I went to find Doreen, who was laying out platters of food—soups and sandwiches, roast chicken and cakes.

"Do you want to see him?" she asked once we'd hugged.

Peter nudged me. "You go with Miranda. I'll stay with Silas; then we'll go up later."

I looked over at Silas, who was standing in a circle of enormous Māori men, all wearing boots and black leather jackets. They didn't seem to mind. And Silas, in his tiny blue raincoat, looked as if he were joining in the conversation.

Taking Miranda's hand, I followed Doreen up to the living room. We took off our shoes at the door. A simple plywood casket sat in the center of the room, surrounded by soft futon mattresses. There were people everywhere, most of them Ngāpuhi, and I was a little nervous about how Miranda would react. I wasn't sure what the cultural protocol was. I hoped she wouldn't start talking about farts. I wasn't sure she even understood what she was looking at.

Skin lay in the open casket, his body surprisingly small and compact. He had his biker jacket on, and his favorite hunting knife was strapped to his belt. There was a patch on his jacket that read "Born to Ride." I took Miranda's hand and knelt beside him. I opened my mouth to explain what was happening, but before I could say anything, Miranda's voice rang out loud and clear in the crowded room.

"You were a good teacher, Skin. But you're dead now." Then, fearlessly, she bent down and kissed him on his grizzled beard.

I swallowed, horrified. But all around us, people started to laugh. Behind me, Doreen patted Miranda on the head. "That's right, darling," she murmured. "That's well said."

Doreen pointed to a collection of Magic Markers scattered around the room. "They're for you to write your good-byes," she encouraged. "On the casket. The little ones, too. Skin loved the little ones."

I looked at what had already been written there: "Catch the big one in heaven, mate" and "Skin, you always took care of our Whānau.** Now I'll take care of yours." Miranda scribbled some drawings on one end, and I took a black pen and wrote, "Skin, I was lucky to know you."

When I looked up, I realized the room was full of children. The casket was covered with kids' drawings. My eyes blurred, and I noticed the cross at Skin's head. "Isaac Shayne Anderson," the marker read. His name wasn't Skin, and it wasn't Dennis. He'd been playing with us all along.

Two days later, we buried him. The church funeral was in the middle of a workday, and Peter couldn't leave the office. But Silas didn't even have school that day. His teachers were going to the service.

Becca and I squeezed into a tight parking place and coaxed the children out of the car. We made our way to the tiny cemetery at the top of the hill, with the century-old white chapel at its center.

I looked around at the carefully ironed shirts, the flushed cheeks and fidgeting children, and after a while I realized that every single person we knew in Purua was in attendance. Nick and Amanda

---

** *Whānau* is an important word in Te Reo Māori. It means "family" in every sense: physical, emotional, and spiritual.

were there with all their daughters, each one turned out in a new skirt and Mary Janes. Autumn and Patrice stood together with their kids, and Bill and Sophia. Then there were the pirates: Skin's best mates and cousins, guys who wore motorcycle jackets and Jim Beam T-shirts, with wide gaps in their teeth when they grinned.

The professional farmers had turned out, too—the tall, grumpy men who worked all day and never slept enough. There were Hamish, John, Dave, and Graham—men who worked huge farms with hundreds of stock. They almost never left their animals, their fencing, or their land.

Silas was at the far end of the cemetery, looking for ways to escape. He was running on a freshly filled grave, watching his shadow dance on the grass. I went down to fetch him, but when I took his hand, he squealed as if I'd slapped him. "No! No! No!" he protested. "Home. Bus. *Away!*"

"*Silas*," I hissed. "You've got to come away from there!"

I dragged him away from the grave, and he let out an ear-piercing screech. Mourners began proceeding toward Skin's grave site, and six young men in black bore the casket on their shoulders. Silas didn't care about any of this. He sat on the ground and refused to get up, and each time I yanked him, he only screeched louder.

Rebecca hurried over with Miranda. "Is everything okay?" she asked, worried. "Do you need me to sit with him?"

"I can't make him calm down," I said finally. "We should go. This isn't fair to anyone. Can you take Miranda?" She nodded, and I reached down and lifted Silas. He continued shrieking, kicking at my legs and trying to bite my shoulder. He looked like a tiny kidnapping victim.

We crept out of the cemetery in disgrace. My cheeks burned, thinking how wild my son was. I was fixated on getting back to the car, pumping up the air-conditioning, and getting out of there, getting back home where no one could hear my kid screaming.

Once we left the cemetery, Silas began to calm down. We made it back to the car and buckled up the kids. A milk truck passed, headed for the highway, and then I started the ignition and pulled onto the narrow dirt road. The service had only just started, so the street was still lined on both sides with cars, a narrow path in the middle our only escape hatch.

We drove past the cemetery and down the road. The milk truck's red brake lights flicked on. "What the . . ." I frowned. Then I laughed, a quick, sharp burst.

"Oh God," I moaned. "We're stuck."

I got out of the car and walked up to the truck, leaving Becca and the children behind.

"Can't move," the truck driver grumbled, indicating the road before him. Large, dusty pickup trucks were parked on both sides, and while the passage between them was wide enough for a small car, there was no way his truck could squeeze through.

"Is there a back road?" I asked. "Maybe I could reverse and go the other way."

"Could do," the driver conceded. "'Cept there's a bloody huge logging truck coming up behind me. He'll be here in two minutes."

I headed back to the car, where Miranda had grabbed one of Silas's books, and he was now beating her over the head with the other one. "Mama!" she shrieked when she saw me. "Silas is going to dead me!"

Rebecca smiled thinly, reaching for her ear buds and turning on her iPod. I sat back in the driver's seat and turned up the air-conditioning. The massive chrome grill of a logging truck took over the whole of our rearview mirror.

"That's it," I sighed. "We're not going anywhere."

I settled back in the parental misery of being stuck in a car with two warring children. I counted to a hundred, trying my best to tune out the screams from the backseat. Vehicles hemmed us in on all sides, before us a steadily narrowing corridor of farmers' shiny SUVs, beat-up family vans, motorcycles, and pickup trucks—the entire motor population of Purua.

The service was blessedly short. After no more than twenty minutes, the congregants began to disperse. The traffic jam was obvious, and as people began to notice it, they hurried to their cars.

I rolled down my window as Amanda walked by. "What are you doing?" she asked, bending down to talk.

"Silas was freaking out," I explained. "We didn't want to bother anyone."

"Nobody cares about that," she scoffed. "It doesn't matter. There're children everywhere. You should have stayed."

"Really? I felt bad."

"Well, you shouldn't do. We all love Silas. Don't we, Silas?" She winked at him.

"Bus," Silas replied.

"I'm having a dinner tonight," Amanda said. "It won't be a raucous thing. I just thought we could get together, eat some food, and remember some good things about Skin."

"We'll be there," I told her. She took my hand and squeezed it. As if by magic, the road ahead was clear. I grinned and started the ignition.

That night when we got to Amanda's, she was slicing a leg of lamb and the wine was already flowing. Silas and Miranda went off to the living room with the other children, swiping thick slabs of bread and butter on the way.

Abi sat next to the wall, barricaded behind a large bowl of salad. "Why did we like him so much?" she wondered out loud.

"Well, he'd have saved you from the zombie apocalypse, for one!" Amanda joked, passing around a platter of lamb. "Skin had a lot of skills."

Nick helped himself to the meat. "I'm a little offended that you all want him to defend you. You can spar with me anytime. I'll teach you all the self-defense you need."

Zane burst out laughing. "Oh man, I'd like to see that! Nick versus Zombie!"

"It wasn't zombies anyway," I murmured. "He got done in by his own heart."

Peter, who had been quiet until now, spoke up. "You know what I liked about him? The guy did what he wanted. Like when he came to our house that day, to roast sheep on a spit. Nine or ten hours, he was just sitting there turning this meat. And I said, 'Don't you have to work?' And he goes, 'Nah. Can't work today. I'm roasting a sheep.'"

"There was a pastor at the wake," Amanda mused. "It was interesting. I mean, not the Jesus part, but what he said was 'If I ask a woman what's most important, she says the husband. If the husband is gone, she says the children. If the children are gone, she says the house.'"

"The house?" I asked, skeptical. "I'd think it would be the rest of the family."

"No." Amanda smiled. "It's the house, because he said the dreams of the family—the *moemoeā*—are in the walls."

There was a silence while we gathered this in. "Great," I commented. "We've got two more months in our house, then we're out. And I'm gonna lose my dreams, too?"

"Nah." Amanda turned her fierce eyebrows on me. "Pitch a tent on our lawn. She'll be right."

I felt my cheeks get hot then, at her easy generosity. I still didn't know what we'd do with the farm animals, but at least my panic about having shelter started to ease. "Thanks," I told her, embarrassed. "You guys have done so much for us. And Autumn, too. We'd be a mess if it weren't for you—"

Amanda cut me off. "Don't be stupid. You'd do it for us."

"I'm just very concerned," Autumn interjected, "about this zombie apocalypse. Skin had all the survival skills. We're stuffed now. Who'll protect our families?"

"We'll have to take care of each other," Amanda said softly. "Now we're the family."

CHAPTER SIXTEEN

THE BINGLEE-DOO

At the start of December, the broody hen got tired of the responsibilities of motherhood and abandoned her eggs, leaving me with a clutch of half-formed chicken abortions.

"What am I supposed to do with these things?" I asked Peter, showing him the pile of cold eggs. They looked normal on the outside, but she'd been sitting on them for two weeks. I shuddered to think what was curled up inside.

"Feed 'em to the dog," Peter suggested matter-of-factly. "That's good meat. It's like chicken veal. She'll love it."

Gingerly, I cracked an egg over Kowhai's food bowl, my eyes screwed shut. The ensuing sound was not what you'd expect when cracking an egg, which is nothing. Instead, I heard a distinct *thunk*.

"Oh, God, this is *so gross*." I moaned, taking the stainless steel dog bowl out to the back deck. Kowhai bounded along behind me, her tongue lolling.

Not wanting anyone to make an omelet with these ghastly things, I prominently labeled the box of eggs, "Chicken Foetuses for Kowhai. Do Not Eat." That second sentence was probably unnecessary, but the eggs didn't really seem to trouble Peter. The situation called for specifics.

In fact, the box of aborted chicks in our refrigerator seemed an apt metaphor for country life, at least as far as we were concerned. Here we'd thought we had a new life in Purua, and now it was coming to an end. We had just one month left in our rented home and no plan for another place to live. Since the two houses we'd looked at had fallen through, no other possibilities had come our way.

This had led to a series of sleepless nights for me, wherein I was awakened by brutal questions such as *What the fuck do I do with the goats?* Like it or not, we were now the custodians of nineteen animals. I could bring myself to plan for some of them. The chickens and sheep could be slaughtered. Anyone would take a healthy cow, because they're worth money. And I felt sure we could pawn off the racist alpacas on Sophia, who found them elegant and didn't know about their fighting teeth.

But I had a soft spot for those baby goats. Not only had I watched them come into the world, but I'd knocked up their mother, put up with her pregnant moods and bad behavior for five months, and brought her special treats when she had cravings. I felt like a second parent to those kids. At the very least, I felt like their baby daddy.

The problem was that Moxie and Stripe were horrifically ill-behaved. I was the only one who found their car-jumping antics hilarious. Guests had started parking at the top of the driveway and walking five minutes to our front porch rather than risk getting their paint dinged up with hoof marks. If permitted to roam, these goats worked like lawn mowers, chewing up every plant and vegetable in their path: the roses, the vegetable garden, the bark from native trees. And when plants weren't enough, they started in on people, sucking on our hair and nibbling at our shirtsleeves.

"Premal will take them," Peter suggested one evening. "That guy from work, the one who eats colostrum. He says goat is fantastic. And now's the time to slaughter, when they're young and tender."

"You're kidding, right?"

Peter rolled his eyes. "We're not starting this again, are we? You can't put them in a sling and lick their assholes. That's got to stop."

"Sure," I conceded. "But I really don't want to eat them. I want to stay out here in the country, but somewhere with proper fencing, and milk them. For cheese."

It sounds simple enough, but "proper fencing" for goats means building a maximum-security prison on your farm. Goats are tireless escape artists. You can try to restrain them with electric fences, but they have the creepy superpower of knowing the moment the power to your fence goes out. Then they'll jump right over it or blast on through. If they're not sure, the mother goat will sometimes push her babies onto the fence, as a heartless science experiment, to see if they get shocked. And if they do get zapped, goats don't jump backward like every other barnyard animal. They jump forward. So, essentially, the electric fence acts as a catapult projecting them on to freedom.

The only truly reliable alternative is goat fencing, which looks more or less like the walls of a prison camp. Eight-foot posts enclose lengths of wire mesh, which are then sunk into the ground, because goats can also dig. We asked a couple of farmers how much it would cost to build some, but as usual, they just shook their heads.

"Aw, ya don't want goat fencing," Hamish corrected me.

"Sure I do," I protested.

"Too dear. Stick to the llamas."

"*Alpacas*," I corrected him. "And what do you mean 'it's too expensive'? How much?"

"A lot," Hamish replied, and took off again on his quad, leaving me to wonder, *How much is a lot? Five hundred dollars? One million dollars?* The answer remained mysterious.

Then, on top of all those worries, Silas was now beating people up. His epilepsy medicine had a number of known side effects, including "increased aggression" and "hyperactivity." In practice, this meant that when he was frustrated, he didn't just moan and call it a day. Instead, he hit me in the face.

One night before Christmas, Autumn and Patrice came over for dinner. I was heading back to the kitchen to grab the ice cream, Silas trotting along behind me.

"Ice keem!" he was chanting. "Ice keem!"

"You'll get your ice cream," I told him. "Go back to the table and wait, please."

Silas found this answer subpar. He grabbed for my thigh and started clawing at the skin. I returned to the table with ice cream, bowls, and a pair of red welts down my legs.

"I have noticed," Patrice observed, "that he is sometimes more aggressive at school. He is hitting, sometimes biting. Is the drug doing this?"

"Yep." I nodded. "It's a known side effect. But what are we gonna do? We can't let him seize all the time."

"Of course."

"What do you do when he hits?" I asked. "Do you punish him?"

"I put him in the cock room."

"The *cock* room?" Peter repeated.

"The *cloakroom*," Autumn corrected.

Finally, in the middle of December, an appointment opened up for an EEG. I imagined that the brain waves would paint a picture, and then all the doctors would gather round. They'd stroke their chins, look thoughtful, and murmur, "Ah, yes. Of course. The binglee-doo. We can fix this."

"I hope you're not getting your hopes up too high," Peter cautioned, seeing the glint in my eye as I packed Silas's bag for the hospital. "It's not like they're going to see a giant bus on the screen, then reach in, pull it out, and the kid'll be quoting Shakespeare."

"Bus," Silas argued.

"I'm not looking for Shakespeare," I sighed. "I just want to learn more."

Silas needed to be asleep for the technicians to get an accurate scan of his brain, so the first thing they did when we got to the hospital was lead us to a private room. A nurse wearing green scrubs the color of toothpaste dimmed the lights, and once she'd taken Silas's vitals, she offered us a sedative to help him sleep.

I took off my shoes and curled up on the bed with my son, grateful for a rest. I'd been up at intervals all night, turning a new batch of Camembert and wondering why I was slave to a pile of milk curd. I was just starting to doze off, Silas's head tucked in the crook of my arm, when someone punched me in the nose.

"Ow!" I squawked, sitting up. "What the—?" Silas was kneeling on the bed, a massive grin spread across his face.

"Bus!" he exclaimed, then kicked me in the pelvis.

"Silas!" I cried, holding both his arms in mine. "No hit! That's mean!"

Silas just laughed, then wiggled out of my grip, slid off the bed, and launched his little body at me, pummeling at me with both his fists. It actually hurt, even though he was five years old and weighed no more than forty pounds. I bent down to talk to him, and he slapped me in the face.

"Arrrgh!" I choked out a strangled cry, then forced myself to count to ten. This hospital room was festooned with signs featuring slogans such as "This Is a No-Hitting Place," and "Violence Is Never Okay." At that particular moment, I disagreed. Actually I think my son could have done with a good cuff on the back of the head, but I held myself back.

When the nurse came in to check on us, I was cornered in the darkened room, both hands flat in front of me to ward off the blows from my homicidal first-grader.

"Help," I gasped. "He won't stop hitting me."

She sat on the edge of the bed. "Is this normal behavior for him?" she asked placidly.

What kind of a fucking question is that? I wanted to snap. *This child is the Antichrist.* But instead, I blurted, "I have no idea. This is not my son."

Of course, I meant this figuratively, as in "This behavior is so unusual that it does not in any way resemble 'normal' for my son," but I think I freaked out the nurse. Immediately, her eyes widened, and she took a fairly hostile tone when she asked, "Then what are you doing here with Cyrus?"

"Silas," I corrected. Then I backtracked. "I mean, he is my son, but this is very strange behavior. What did you give him for a sedative? Methamphetamine? He's acting psychotic." Silas was now slamming his little body into mine, trying to take me out at the knees.

The nurse seemed reassured, though she made no move to

protect me. "I'll get the doctor," she said. "Sometimes, with these sorts of children, the sedative can have an opposite effect. We'll try an alternate drug."

Then she left me with the midget psychopath, who was hollering "*Out! Out! Away!*" and trying to drag me to the exit.

"At least it's not 'bus,'" I muttered, and steeled myself for the next round of drugs.

I coaxed Silas onto the hospital bed, tempting him with a blue latex glove that I blew up like a balloon. I tossed it to him, and he batted it back.

"Good!" I smiled. "Ball. Throw ball!"

"Ball," Silas repeated, the old familiar glint in his eye. He reached for the glove, and then his arm just kept reaching. Both arms shot out, rigid. And then he was tipping backward.

I lunged for the call device, hitting it four times in a sharp staccato. A nurse came in, her mouth open to speak. Then she was at the bed in two quick strides and hitting the call button for backup.

"How long has it been?" she barked at me.

"Fifteen seconds," I stammered. "Twenty. This is the worst one I've seen."

Silas jerked involuntarily, as though his small body were caught in the mouth of an invisible dog, shaking him and shaking him. The skin around his lips went from gray to blue. Two more nurses entered, but there wasn't anything they could do. One held him down; the other scanned her watch.

"Time?" she asked, touching me on the shoulder, because I didn't look up. *Come on*, I urged Silas silently. *Come back, Silas.* "One minute ten," I snapped, glancing up at the clock.

"Someone get the doctor," I heard one of the nurses say. Then a nurse turned and speed-walked out the door.

At two minutes twenty, Silas was still. I leaned over him on the bed, smoothing his hair and kissing his cheek, which was cool and damp. The doctor walked in then, this one a balding, slightly befuddled young resident. He asked for a description of the seizure and took notes, then looked down at Silas and awkwardly patted his knee.

"Well, the good news is, he's sound asleep now," he observed. "We won't need a sedative to do the EEG."

The door swung open and the technician moved in with her rolling computer. She leaned over, marking small dots on Silas's head where the electrodes should go.

I sat there, my book forgotten on my lap, as she drew on my son. Then she started attaching wires to his scalp. By the time she was finished, Silas looked like electronic spaghetti. Then, to complete the picture, she fit a hairnet over the top.

"Keeps everything in place," she explained.

I never thought I'd be sitting in a New Zealand hospital watching my disabled five-year-old get his brain hooked up to a computer. But, then again, I'd never expected any of this. Didn't think I'd carry goat shit around in my purse or pull hot guts from a freshly killed turkey. Didn't think I'd stay up all night flipping milk curds, or feed chicken abortions to my German shepherd.

Twenty minutes into the procedure, Peter called my cell phone. "How you holding up?" he asked, concerned. "Do you need me?"

"That's okay," I told him. "I got this."

The technician was watching brain waves plod slowly across her monitor, and suddenly they all went nuts. The black lines swung wide, getting so large and dense that her monitor looked black. I watched the technician's face. She leaned in, concentrating.

"How's Silas doing?" Peter was asking.

"I don't know," I told him. "I'll let you know how it goes."

And it was true, I didn't have a clue how Silas was. The technician wasn't allowed to tell me a thing, because we had to wait for the doctor's report. When she finished the EEG, they released us from the hospital, and we didn't know anything more than before.

We had a new prescription that raised Silas's medicine dose again, and the new drug levels made him stumble and fall. A few days later, Silas wandered out of bed in his chocolate-brown moose pajamas, his tousled blond hair sticking out in all directions.

"Baaaaahh . . ." he moaned at me, his eyes glassy. "Baaaaa . . ."

"Mama?" Miranda asked.

"He sounds like a ghost," I observed, watching as Silas staggered into a chair.

"I think he's trying to say 'bus,'" Peter reasoned. "But it's falling apart in his mouth."

"Mama?" Miranda asked again. Her chin started to tremble.

"What, Magnolia?" I looked down at her. She was still rumpled from sleep, in her pink nightie with rabbits. In one hand, she held her gold beaded evening purse. She opened it up and pulled out a rock to show me.

"Sometimes I throw a rock at my face."

"You do?"

"She's worried about her brother, I'll bet." Peter finished his coffee. "That's what the rock's about."

Silas seized an empty coffee cup and tipped it in his mouth, smacking his lips at the sugary dregs. "I know," I said softly. "We're all worried."

The phone rang. I went to pick it up, and when I heard the mellow Ghanaian accent on the other end, I put my hand over the receiver and hissed to Peter, "Silas's doctor."

I went out to the laundry room so I could talk without being disturbed. "We have the results of the EEG," Dr. Osei announced, without any preliminaries. "The neurologists and I have had a chance to review them, and there is evidence of seizure activity."

I waited. "Yes?"

"The EEG shows that Silas is having seizures."

"But we already knew that."

"At this stage, there's no other useful information on the report. We don't know what is causing the seizures, and we don't know where they're coming from."

I thanked the doctor and hung up, trying hard to control my temper. Peter opened the door behind me. "What did he say?"

"That Silas is having seizures."

"Well, he sure as shit wasn't tap-dancing. That's all?"

"Yep, that's all. I'm going out to feed the animals."

I went out through the winery, noting with mild despair that the strawberry wine had back-filled the airlock with a lurid red pulp, so the wine seemed topped with boiling guts. I checked my latest cheeses in the pantry, four rounds of chèvre with a blossoming white rind on them, already sporting a five o'clock shadow of cat hair. But none of it really mattered. We'd have to leave this house anyway, right after Christmas. Rebecca would go back to the States, and we'd be selling or giving away all the animals. We'd be living in a tent soon enough, or else back in town, with one kid stumbling and groaning like a drunk and the other one hitting herself in the face with a rock.

I heaved a self-pitying sigh. Inside the house, the phone started ringing again. Peter picked it up and said something, then he rolled back the sliding glass door and handed me the receiver. "It's Fiona."

"Fiona? I don't know anyone named Fiona. Tell her I'm not here." I just wanted to hang out with my goats, scratch their heads, and apologize for giving them away.

Peter listened for a minute, then put his hand on the receiver. "I think you might want to talk to her. She says she's selling her house."

CHAPTER SEVENTEEN

NO CROCODILES HERE

I hung up the phone. "Four bedrooms!" I squealed to Peter. "Twelve acres! It's just down the road! There're trees! And a greenhouse! And there's room for the goats!"

"Sugar, Honey, Ice, and Tea!" Rebecca squealed. "That's amazing!"

"Settle down," Peter cautioned. He was always more careful than I, especially when it came to money.

We showed up at Fiona's house that Saturday, bringing one drunken boy on epilepsy medication and a little girl in a rainbow bathing suit with a blue feather handbag. The driveway twisted back from the road, lined with puriri, totara, and other native trees. Peter and I sat up straight in our seats, both leaning slightly to the left to catch the first glimpse of the house.

"Maybe it'll be a farmhouse," I speculated. "One of the really old ones, made with kauri wood, with those beefy floorboards and the old-fashioned windows with the wavy glass." It wasn't impossible.

Out in the country, there were still a few of those nineteenth-century farmhouses left. "Maybe it'll be—oh."

At first I hated it.

Rising up from the gravel driveway was a split-level home with dark gray weatherboards and concrete facing on the brickwork to make it resemble stone. It seemed so conventional, like somewhere you'd live if you moved to the suburbs.

But, then again, a low rock wall curved around the house, rippled with moss and young ferns. Fiona and her husband, Dave, stood awkwardly in their driveway, waiting to greet us. We got out and introduced ourselves.

"No! No! No!" Silas protested. New faces made him nervous.

"That's all right," I soothed, handing him his Dart. "You can stay in the car."

Miranda walked up to Fiona, a fine-boned woman in a gauzy summer blouse, and demanded, "Are there crocodiles here?"

"No," Fiona said, taken aback. "I don't think there are any crocodiles in New Zealand, actually."

"That's good." Miranda turned to me. "We can buy this house, Mama. Because there are no crocodiles."

"Right, then," Dave interjected. "Now we've sorted that bit, let us show you around."

They walked us through established vegetable gardens bursting with tomatoes, cucumbers, corn, and chard, up a winding stone passage to the flower gardens, which Fiona had planted with an array of colorful blooms. A porch wrapped around two sides of the house, looking out over stands of gum trees and distant hills.

"Five paddocks," Dave told Peter. "All with electric fencing. We run beef now, but you could keep sheep, goats, what's that you've got? Camels?"

"Alpacas," Peter clarified.

"Right, that. They'll be happy here."

"And a henhouse," Fiona pointed out.

"Do you have running water?" I asked tentatively. "Cold *and* hot?"

Dave turned around and looked at me inquiringly. "Yep, 'course. And separate water from the creek for the animals."

"What about power?" Peter wanted to know. "Like electricity, that comes out of the walls?"

Fiona and Dave exchanged a look. "Why wouldn't we?" Fiona asked. "We are in New Zealand. It's not the bloody third world."

"I know." Peter breathed a sigh of relief. "It's just . . . we've seen a few rough places."

"Come on." Dave grinned. "Let's show you the rest of the property."

I went out back to check on Silas, who seemed content with his Dart and his books. I made sure the windows were open and he had enough shade, then I strolled back inside. Dave was showing Peter and Rebecca the garage, an open space with a broad, sturdy workbench in the back.

"Perfect!" I crooned. "My lair!"

"Pardon?" Fiona asked.

"She's into making wine and cheese," Peter explained, looking slightly embarrassed.

"Right here," I announced, looking around with satisfaction. "In my fermentation lair."

"But you'll keep a car in here, too," Dave clarified.

"No way," I told him. "Just wine and cheese. Also, I might start making sausages."

Dave looked as if he wanted to ask more questions but then seemed to think it best to move on. They guided us outside and brought us through the orchard, where peach, plum, and apple trees grew among lemon and mandarin orange trees. There was a wide, well-fenced chicken run, with a solid henhouse, draped in bright green passion fruit vines. We walked to the top of the hill, past massive maples and oaks, and down the other side, to a small creek and an area they'd fenced off for native bush.

By this point, I was entirely captivated. Every time Fiona turned her back, I gesticulated wildly at Peter, mouthing, "Yes. *Yes*. *Let's get it. Yes*."

Peter drew me close and hissed in my ear, "*Shut up*. Don't act happy."

"Pardon?" Dave turned around.

"Nothing," Peter said, covering.

As they walked us back to our car, I pointed to a tall tree with round, glossy leaves and dark pink blooms. "What's that?" I asked Fiona. "It's stunning."

"That's my magnolia," she told me, smiling. "I'll miss it when I go."

"See?" I squeezed Miranda's hand. "There's a magnolia here already."

"Yep." Fiona nodded. "I love that one. It's probably my favorite. That and the kowhai."

By the end of that day, we had a deal. Fortified with three glasses of strawberry wine, Peter called Fiona and hammered out a price, getting her to throw in the greenhouse, the spa pool, the ride-on lawn mower, and their Land Rover with the deal.

"What can I say?" Peter said modestly, hanging up the phone. "I'm just a fantastic businessman."

Suddenly, we had three weeks to move into our new house. The days became a blur of paperwork and bank appointments, as we maneuvered through the bureaucracy that comes with buying land. Packing our boxes was easy enough, especially since Rebecca was there to help out. Still, even though Katya and Derek were reclaiming their cat and dog, we were still faced with the challenge of moving nineteen farm animals down the road to a new place.

"I guess I could hire an animal transport," I moaned to Amanda one night. "But that seems so expensive and silly. Our new house is just three kilometers away."

I'd brought several bottles of strawberry wine. This drink looked like Kool-Aid and tasted like jam, but when you swirled it, you saw it had legs like a Cognac. Amanda was on her third glass.

"Just walk them," she told me recklessly. "There's no need for a truck."

"*Walk them?*" I repeated. "Three alpacas, two cows, two sheep, three goats, six chickens, one rooster, a dog, and a cat? What, do I bring a net?"

"It'll be fun," she insisted. "We'll get the whole community involved."

More strawberry wine flowed, and this began to seem like a clever idea. We'd put all the animals on halters, and the children could help us lead them. Becca could help us weave flower garlands for their necks, and we'd paint their horns in shimmering gold.

"Renaissance minstrels!" I insisted, toward the end of the third bottle. "We need recorders, tambourines, and a lute!"

"What's a lute?"

"It's a harpy thing. With strings. You can't do an animal parade without a lute."

So that night, when we got home, Rebecca and I tapped out an e-mail to the Auckland Philharmonic:

> Dear sirs,
>
> We are seeking a small group of musicians familiar with Renaissance ballads to accompany us on a Parade of the Animals in Purua next month. Recorder and lute required; trumpet, viola and guitar optional. Must be a fan of goats.
>
> Please let me know if your musicians are free.
>
> Kind regards,
>
> Antonia Murphy

The next morning, I woke up in my bed still wearing my clothes and uncertain where my children were. "Oh, no," I groaned, grabbing a pillow and pressing it over my face.

"Headache?" Peter asked, amused.

"I think I drunk e-mailed the Auckland Philharmonic last night."

"Ah." Peter grinned. "Yes, I think you did. But don't worry. I'd be shocked if you heard anything back."

And, curiously, we didn't.

So we decided to go with plan B, which was asking local farmers for assistance. Maria's husband, John, came round a week before we moved, to check out our animals and tell us if he could help.

"Yep, the alpacas and goats are no problem," he assured us, hands shoved deep in the pockets of his shorts. "I've got my trailer, with a cage on top. We can ride 'em in there." He rocked back and forth on his gumboots, thinking. "But your lambs have got flystrike."

"What?" Becca stepped in to scratch her sheep between his horns. "Not Ba. He's perfect."

"Yep." John peered at the sheep's backside. "See how he's holding his head down? And he's real daggy there in the back? You'll have to crutch him."

"I can't do that," I said, feeling ill. "I don't do sheep butt."

"There was a time—" Peter interjected, but I stepped on his foot to shut him up.

"What's wrong with Ba?" Miranda asked John, her eyes wide with worry.

"Nothing, Magnolia," I told her. "He just has worms in his bottom. It's nothing."

"Cool," Miranda said. "Do they suck the blood?"

"Nah," John corrected her. "They eat the meat, more like. You let it go too long, they'll eat yer sheep all up."

"Wow." Miranda breathed, impressed.

"It's no worries," John told me, peering over the top of his sunglasses. "I'll do it when I move him. It's a two-second job. Not a problem."

"Really? Thank you." John scowled when I spoke, and I realized there was a time when I would have thought he was angry at me. But this was a man who'd offered to shave a sheep's ass for free. What's a better indication of friendship than that?

We moved on a bright, clear day in January, and the process went smoothly, despite our lack of lutes. In just a few days, we were settled at home. The children shared a bunk bed in one bedroom, with Rebecca in her own room across the hall. Peter and I took over the master bedroom. We all had lush, leafy treetops visible from each of our windows. Each group of animals had their very own paddock, with fencing that actually worked. Peter built a little house for the goats, and on quiet mornings we could hear them clattering happily on the metal roof. The alpacas had plenty of grass and shade, and the cows and sheep kept one another company in the lower paddock, where they could rest easy without danger of attack. Kowhai didn't have a bed, so she dug a hole in the flower garden and flopped gratefully on the cool dirt inside.

A week after we moved in, it was time for Rebecca to say goodbye. "I can't believe this!" I moaned. "Who's going to help take care of the animals? Who's going to teach me how to weave?"

"Who's going to hand you tissues when you need to take a shit in Ruatangata?" Peter chimed in.

"I could just stay. Drop out of school. I don't even want to go back," Rebecca offered.

"*No way.*" Peter went for his car keys. "Your mother would kill me. I'm driving you to the airport right now."

That evening, we sat out on our wraparound deck. We were sipping a strawberry-rhubarb wine, a special blend I'd concocted that looked like a sunset and tasted like pie. "I didn't think it would be so important," I said to Peter, "having a house of our own, in Purua. It feels like we're finally home."

"Yeah." Peter reached for my hand. "Now we just have to get on top of that garden."

It seems odd, looking back, that after a year in the country, I hadn't really done any gardening. Now we didn't have a choice. Fiona had left us her late-summer vegetable garden: six beds overflowing with tangled tomato vines, eight-foot sunflowers, and tall, woody stalks of red chard.

As usual when confronted with the impossible, I called in Autumn for help. "I kind of like it wild-looking," I confided, admiring the looming sunflowers, their blossoms the size of satellite dishes.

But Autumn rolled her eyes. "You've got noxious weeds here.

You can't just let them go. They'll take out your garden, kill all your vegetables."

"Wait a minute," I countered. "What's noxious?"

"Creeping oxalis. Deadly nightshade. Ladder fern."

"Ha!" I crowed triumphantly. "I grant you that deadly nightshade doesn't sound like something I'd eat, but what's wrong with ladder ferns? They sound kind of cute."

"They'll get in that pretty rock wall you've got and pull out the rocks till the whole thing falls down."

Her argument was persuasive, and I agreed that I might need some help.

Autumn had some free time the following weekend, and she promised to help me get my garden under control. For the next few days I could have been unpacking boxes and starting the weeding, but instead, I pondered my gardening outfit. I imagined myself in flowing linens, perhaps with a kerchief, backlit in a sunflower glow. I'd be just like a Van Gogh peasant, except not so ass-heavy, and minus the freaky yellow sky.

But no sooner had I tied my kerchief that Saturday than Autumn stuck a tool in my hand. "Here," she told me, "dig." I examined the implement. Long stick, sharp metal bit at the end. Not a shovel, so . . . a scimitar? Quarterstaff? Mace?

"It's a hoe," Autumn informed me. "You turn the soil and dig out the weeds with it."

I considered telling her that where I come from, a ho is a loose woman, sometimes also known as a "cheap ho" or a "skanky-ass ho," but that just seemed rude. Instead, I started to dig.

While we were working, I tried being chatty. "So," I started brightly, wiping the sweat from my eyes, "are there, like, shortcuts in the garden? Ways to not work so hard, maybe? Staff we could hire?"

Autumn grunted as she hauled out earth with a spade, turning it and breaking it up as she worked. "The key to keeping your garden healthy is to feed the soil," she explained. "So all the poos your animals make, all the goat berries and the cow pies and the rest of it—you work that into your soil."

"I knew we could use it!" I slammed my skanky-ass ho into a huge crop of nightshade. "So goat poo, cow poo, sheep poo, and—"

"Worm poo," Autumn finished for me. "Castings, actually."

"Hold up." I laid down my hoe for a minute. "Where does a person get *worm poo?*"

"A worm farm," she answered, as though this were obvious. "You'll get the castings, which are the poo, and then you can make worm tea."

And that's how I came to find myself the proud owner of a brand-new worm farm, which arrived with a pound of live worms. "Each bag," the accompanying literature boasted, "contains approximately 2,000 live tiger worms."

The Hungry Bin worm farm was formed from three green plastic bins, but its contents were the stuff of horror movies. Tangles of glistening worms slithered through piles of kitchen scraps, eating their way through a slurry of putrid meat, moldering vegetables, and vacuum cleaner lint. Each day, I checked the tray at the bottom for "tea," the cocktail of worm excreta and rancid vegetable juice that's supposed to be a peerless fertilizer. Instead, all I found were maggots. This worried me, so I consulted the manual.

The booklet was surprisingly placid on the subject. "While many people find maggots unpleasant, they will not harm you or your worms. In fact, they are good decomposers and, like worms, will produce a high-quality casting."

This posed a philosophical problem. I reported my findings at dinner that night. "So we feed our old food to the worms and the maggots, who shit out the castings," I began.

Peter pushed his salad away and took a breath, looking patiently in my direction.

"Then we feed the worm poop to the garden, which makes vegetables for us to eat."

He nodded.

"We eat the vegetables, we feed the scraps to the chickens and the goats, we get eggs and goat milk and feed the scraps to the worm farm, and they shit out more castings."

Peter widened his eyes, as if this were obvious.

"So that's all there is?" I slammed my fork down to emphasize the point. "Just shit and worms and maggots and rot? And all we can look forward to is . . . being *cast*?"

"Yep." He sipped his wine. "That's one way of looking at it."

Silas crawled on my lap, and I hugged him, burying my face in his sweaty blond hair. "So all the fun parts in the middle? The books and the sunsets and the wine? Dinner parties and friends and sailing trips in the South Pacific? That's just a stop on the way to the worms?"

"And tomatoes, Mama!" Miranda held up a tomato, from which she had sucked all the juice. It sagged in her grip. "We also get lots of juicy delicious tomatoes!"

Peter shrugged his shoulders. "The worm's a perfect being, really. It's a tube that eats and poops and makes more worms."

I considered this for a moment. Silas wiggled on my lap. "Kiss," he said, and I listened, kissing the top of his head.

It was gruesome but sort of comforting, this rule of the worm. Each day, when I opened the worm farm and tipped in coffee grounds and eggshells, half-eaten sandwiches and orange rinds, their slithering tangles reminded me of where we would all end up. And until we're all cast, as Miranda would say, we get to suck the delicious tomatoes.

Miranda helped me pick whole baskets of tomatoes, and together we made liters of sauce, so we could taste their tangy sweetness year round. I learned how to identify nightshade and kill it, and Peter constructed a compost bin from old pallets. We drank rhubarb wine when Peter got home at night, and we strolled through the fields with a wheelbarrow, collecting manure for the garden.

Silas still staggered around with a faraway look in his eye, and once the move was over, I screwed up my nerve and wrote to his doctors. "I want to have a serious discussion about taking him off this medicine," I told them. "My son is drooling and falling. It's not like him."

"Sometimes these side effects can happen with these sorts of children," the doctors wrote. "It may take us a while to adjust the dose."

Reading this, I hit the Reply button. "Be that as it may," I wrote, "I know my son. He might not talk much, but he doesn't drool. And he doesn't stumble and fall. I want him off those meds."

And they listened. The neurologist wrote back, suggesting a schedule to wean him off his medicine. The first day we cut down his dose, he swiped his sister's DVD and clocked her over the head with it. Then, when I yelled at him, he hopped up and down, laughing.

"There's my boy," I told Peter, my heart full of love and relief. "That's the midget ghoul I remember."

With Peter away at work most of the day, I was in charge of the farm. I thought I'd be a hot mess without Rebecca there to help out, but caring for all those animals wasn't as hard as I'd expected. With irrigated water troughs and sturdy electric fencing, they didn't really need much care. They strolled around contentedly, munching on grass and sunbathing in their wide-open paddocks.

It was starting to feel a little too easy. "What do you think about getting more animals?" I asked Peter one day. "We have all this land to keep them on."

"No pigs," Peter said emphatically, shuddering as he glanced down at his boots.

"I wasn't thinking about pigs," I replied. "We don't have enough food to feed them, anyway, at least not until we can give them free milk from our cows." I scrolled through the poultry section in the online farm listings, clicking on one that looked good. "How about ducks?"

"You mean *rapists*? Are you kidding me? What about the chickens?"

"This lady says her ducks are gentle. And they're Muskovies. So they'll be good for making pâté." *And confit*, I thought with delight. Confit is a delicious French health food in which meat is poached slowly in its own fat. My new fermentation lair had room for all kinds of medieval food projects. I might even try a DIY foie gras.

So on Saturday I drove out to pick up the ducks and came back with a mated pair in the trunk of my car. These Muskovies were surprisingly elegant birds. Their feathers weren't white but linen, oyster, and ecru, shades as subtly nuanced as a high-fashion bridal gown. And since they had each other to mate with, I didn't think they'd bother my hens.

But they did make us nervous. To begin with, these ducks had claws. "You'll have to hold them close when you pick them up," the lady told me as she loaded them into my car. "Else they'll scratch you like a feral cat."

And then there was the matter of the quack. Muskovies, I soon learned, don't. No cute little quacking bird sounds for these ducks. Instead, when they came over to greet us, they hissed like venomous snakes. Baby ducklings aside, I wasn't sure I'd done the right thing by adopting them.

"What should we call them?" I asked Miranda. "I was thinking Confit and Pâté."

"No, that's not their names," she informed me. She was wearing her favorite Snow White dress; these days she never took it off. The yellow tulle skirt was tattered and stained.

"It's not? What are their names, then?"

"That big one," Miranda pointed to the drake, "is Daddy Yankee. And the girl one is Nicki Minaj." Though Rebecca was gone, she'd left her mark. Our ducks would be named after a Puerto Rican pop star and a hip-hop diva from Queens.

Hearing his new name, Daddy Yankee waddled over to Miranda, hissing and puffing his feathers. I hustled Miranda out of the chicken run before he could try any funny business.

But the ducks seemed placid enough, and it was time to start planning a housewarming party. "What are we going to serve?" Peter asked. "You can't do sheep on a spit."

Skin had been the designated sheep roaster of Purua, and it felt too soon to replace him. "That's all right," I said. "We'll do something American. Ham glazed in Coca-Cola. We'll freak them all out."

The day of the party was gray and humid, with scattered showers all afternoon. But most of our friends turned out, rolling up the curving driveway and parking in our gravel courtyard. Abi and Zane brought honey from their bees, and a frightening chutney that Abi's mother had made fifteen years ago. Sophia and Bill brought a pretty silk bag containing a selection of gourmet sauces and jams. Maris and Nova brought a calf skull that they'd boiled especially for me and then decorated with twisted forks and peacock feathers. They held it out shyly, the cutlery swinging morbidly from the bones.

"I *love* it," I told them, kissing them both. "Maybe I'll make it into a hat."

Lish arrived with her daughter Amber. It was a warm night, but she had a shawl pulled tight around her shoulders. "How've you been?" I asked, hugging her. "You're not moving away, are you?"

"Pfff." Lish waved her hand dismissively. "I won't lie, it's not been easy. But people ask if I'm leaving, and I say, 'Why would I? I'm not here because of Skin. This is my home.'"

"Yeah," I poured a glass of homemade cider and gave it to her. "Sneaks up on you, doesn't it?"

"I'll give you a day of free gardening!" sang a merry voice, and I turned to see Maria stride in with John, holding up a bowl of her broad beans with fresh dill and feta. John made his way to the corner with the men, nursing a beer and looking stern.

"John?" I asked him shyly. "Do you think you would help me carve the ham?"

John actually looked grateful for something to do, and I followed him out to the kitchen. "Thanks for doing the crutching for Cou-Cou and Ba," I ventured. "Were there really maggots in there?"

"Yep," he said matter-of-factly, not looking up from the ham.

I shuddered, thinking of the ass I'd almost licked.

"Not too bad, though. Meat's still good. Where d'you want this ham?"

I handed him a platter, then asked, "So, when can we . . . you know . . ."

"Yep?"

"Slaughter."

"Ah, that. Could do it now, if you like. For the best yield, I'd wait a few months, May or June."

"You doing sheep on a spit?" Autumn asked, coming to the kitchen for a glass. "Be tough without the master."

"I know," I admitted. "But I can learn. I'll get a book from the library."

The party went late that night, till the cider was gone and the ham was devoured. The last guests to leave were Amanda and Nick, because Nick had Peter in a headlock and wanted to show him how to fight back. They carried out their youngest daughters wrapped in blankets, while Sophie stumbled along behind them.

"Shall we check on the ducks?" I asked Peter, offering him my hand.

"You know, Hamish never showed up," I mentioned, as we made our way outside. "I put an invitation in his mailbox. Maybe he hates me after all."

"No." Peter shook his head. "It was never about you. He's a dairy farmer. He works seven days a week, from four in the morning until night. He's always covered in cow shit. He's just tired."

We strolled out in the moonlight, walking the short distance through the orchard to the henhouse, where the chickens were all fast asleep. And there, in the middle of the chicken run, sat the two ducks, curled around each other for warmth. They looked up curiously when we entered.

"They are kind of cute," Peter admitted.

"And to be fair, Quackers the rapist only turned mean when his mate died," I pointed out. "He was lonely."

"And it was ten years ago."

"Still." I pointed a finger at the two of them. "Try any funny business with my chickens, you two, and I'll kill you and render your fat. Then poach you in your own juices."

The ducks were unimpressed. Daddy Yankee looked up and hissed.

I tried again. "And then I'll feed your bones to the worms."

Nicki Minaj got up, waggled her tail, and pooped a green turd on the ground. Then the two of them waddled to the far end of the chicken run, where they settled down to sleep in the bushes.

"Oh, fine," I snapped. "Walk away. Go to sleep. Make yourself at home, why don't you?"

Peter kissed me on the neck. "Why not?" he asked. "We have."

EPILOGUE

Six weeks after we took Silas off his epilepsy medicine, he was playing quietly in the living room when he had another seizure. This time Peter had to pinch him on the arm to wake him up. We went in to see the pediatrician and discuss putting him on new medication.

"There are two options," Dr. Osei told us, "and both, unfortunately, have side effects. One of the drugs has a sedative effect, and the other one could result in a rash."

"The rash sounds good," Peter commented. "I mean, does it just look bad?"

"Well"—the doctor shrugged his shoulders—"he would need to come to hospital. The sores can burst, you see, and the skin begins to slough off. Then there can be infections."

Apart from his wonderful hugs, nothing is easy with Silas. But given the choice between those two options, we decided to preserve Silas's cognitive abilities as much as possible. We went with the risk of infected sores.

So far, it's worked out fine. No rash, no seizures, no drunken moaning, and Patrice reports that Silas is concentrating better in school.

Miranda, meanwhile, is doing a great job showing us how hard it can be to have a kid who *does* talk. Though not quite four, she has entered a belligerent teenager phase, characterized largely by slamming doors and screaming "*No!*" at the top of her lungs.

On the other hand, she's mostly cute. When she found a dead butterfly in our driveway one day, she decided she wanted to make a bed for it. We put a silk handkerchief in a plastic sandwich box, and I explained that when we make a bed for a dead thing, it's actually called a bier. So now, when we're out in public, Miranda will occasionally panic, tugging on my arm and demanding, "Mama, where is my *bier*?" This has won us some disapproving looks from passersby.

Autumn took over in January as the principal of Purua School, and while I know my kids have a teacher who's sweet, smart, and engaging, it also means we almost never get to see her. It turns out that running a one-room schoolhouse is a lot of work, and she doesn't have much time for strawberry wine.

Sophia hasn't retired, but she did move on to work as assistant principal at the special-needs school in town. She loves the job, and it's comforting to know she's there. If someday Silas does need to enroll at that school, I know he'll be in good hands.

Amanda, Nick, and their family continue to thrive. Nick is teaching self-defense classes in the community, so we'll all be safer when the zombie apocalypse hits.

Zane likes to come over at night and shoot possums with Peter—possums are an aggressive pest in New Zealand, and killing them is considered a national service—but Abi won't have any part of it. She's not a big fan of guns.

Maria's trying to get me to join her amateur rugby league, but I had to explain that I'm a nerd and I don't play games with balls. "Except Peter's," she quipped, and I had to concede the point.

Katya and Derek came back from Germany, and they were unimpressed with the state of their carpet. We wrote them a big check and apologized, and they're starting to forgive us.

I don't see Hamish that often, now that we've moved and we're not over at his house begging for free milk every day, but I do sometimes glimpse him in his paddocks, working his cows. He always waves to me, and sometimes I even get a nod.

Lish's daughter Amber decided to move back home with her mother for now. The first time I met her, she took my breath away. She has her mother's beauty, long wavy hair, and her father's eyes—their kindness and warmth. As far as I can tell, her soul is as gentle as her dad's, and she's training to coach sports for kids, both able-bodied and with special needs.

A month or so after Skin died, Autumn and Amanda went out and bought a hand-carved trophy, made from kauri with a paua shell inlay. It's now the "Skin Anderson Award" for Purua School, given annually to the child who is the hardest worker—and, presumably, who knows how to roast a sheep on a spit.

All our animals are all still with us. Soon after we moved to the new house, Peter built a play structure for Pearl and her babies, a tower some twelve feet high, which they're still trying to conquer. There's a slide for them, too, made from a long sheet of corrugated tin. So much more fun than a car! (And much, much easier on the windshields.)

Peter is still casting about for fantastic business ideas. His latest plan is to buy beehives and sell off all the honey. This should result in chaos, bee stings, and confusion—which is exactly the way we like things around here.

We still love our new home, especially sitting on the deck at sunset, sipping our homemade wine while looking out at the distant hills, the stands of gum trees and the pine forests. My only complaint is that it's so big—with five paddocks and just ten livestock—that we hardly ever see our animals.

So I'm pretty sure we'll be paying a visit to Gay and Mike soon, to stock up on new racist zombie alpacas. I'll have to figure out how to inseminate Cinnamon and Lil' Lady, either by getting a bull up here or (gasp!) donning full-length gloves and doing it myself, artificially. And as for Pearl, Moxie, and Stripe, I think they're about due for a visit to Love Mountain.

> Antonia Murphy
> Purua, Whangarei
> New Zealand
> April 2014

Penguin Random House LLC
1745 Broadway
US-NY, 10019
US
https://www.penguinrandomhouse.com
1-800-733-3000

The authorized representative in the EU for product safety and compliance is

Penguin Random House Ireland
Morrison Chambers, 32 Nassau Street
D02 YH68

https://eu-contact.penguin.ie

ISBN: 9781592409549
Release ID: 150393642

Printed in the United States
by Baker & Taylor Publisher Services